Honda C50 C70 and C90 Owners Workshop Manual

by Mervyn Bleach

Models covered:

C50 49 cc   First introduced into UK February 1967
C70 72 cc   First introduced into UK February 1972
C90 89.5 cc   First introduced into UK September 1967

These models are all step-through scooter style machines, with different engine capacities.

ISBN 0 85696 324 0

© Haynes Publishing Group 1977

ABCDE
FGHIJ
KLMNO
PQ

All rights reserved. No part of this book may be reproduced or transmitted in any form or by any means, electronic or mechanical, including photocopying, recording or by any information storage or retrieval system, without permission in writing from the copyright holder.

Printed in England

HAYNES PUBLISHING GROUP
SPARKFORD YEOVIL SOMERSET ENGLAND
distributed in the USA by
HAYNES PUBLICATIONS INC
861 LAWRENCE DRIVE
NEWBURY PARK
CALIFORNIA 91320
USA

Acknowledgements

Grateful thanks are due to Honda (UK) Limited for the technical assistance given whilst this manual was being prepared and for permission to reproduce their drawings.

The C70 model on which the photographic sequence was based was supplied by Fran Ridewood & Co of Wells.

Brian Horsfall and Martin Penny assisted with the dismantling and reassembly and devised the ingenious methods used for overcoming the lack of service tools.

Les Brazier and Leon Martindale arranged and took the photographs, Jeff Clew edited the text and advised about its presentation.

We would also like to acknowledge the help of the Avon Rubber Company, who kindly supplied the illustrations about tyre fitting, the Champion Sparking Plug Company Limited, for providing the illustrations about plug maintenance and electrode conditions, and Renold Limited, for information about equivalent British - made chains.

Finally, our grateful thanks to Arthur Vincent, of Vincent and Jerrom Ltd., Taunton, for permitting us to photograph the three models shown on the front cover of this manual.

About this manual

The author of this manual has the conviction that the only way in which a meaningful and easy to follow text can be written is first to do the work himself, under conditions similar to those found in the average household. As a result, the hands seen in the photographs are those of the author. Even the machine was not new: an example that had covered a considerable mileage was selected so that the conditions encountered would be typical of those found by the average owner. Unless specially mentioned, and therefore considered essential, Honda service tools have not been used. There is invariably some alternative means of loosening a vital component when service tools are not available but risk of damage should always be avoided.

Each of the seven Chapters is divided into numbered Sections. Within these Sections are numbered paragraphs. Cross reference throughout the manual is quite straightforward and logical. When reference is made 'See Section 6.10' it means Section 6, paragraph 10 in the same Chapter. If another Chapter were meant, the reference would read 'See Chapter 2, Section 6.10'. All paragraphs are captioned with a Section/paragraph number to which they refer, and are relevant to the Chapter text adjacent.

Figures (usually line illustrations) appear in a logical but numerical order, within a given Chapter. Fig. 1.1. therefore refers to the first figure in Chapter 1.

Left-hand and right-hand descriptions of the machines and their components refer to the left and right of a given machine when the rider is seated normally.

Whilst every care is taken to ensure that the information in this manual is correct no liability can be accepted by the author or the publishers for loss, damage or injury caused by any errors in or omissions from the information given.

Contents

Chapter	Section	Page
Introductory sections	Acknowledgements	2
	About this manual	2
	Introduction to the Honda 50, 70 and 90 models	6
	General Machine specifications	6
	Ordering spare parts	6
	Routine maintenance	7
	Recommended lubricants	13
	Working conditions and tools	14
Chapter 1: Engine and gearbox	Specifications	15
	Removal	17
	Dismantling	19
	Components	42
	Reassembly	46
	Fault diagnosis	59
Chapter 2: Clutch	Specifications	61
	Dismantling	61
	Examination and renovation	64
	Reassembly	65
	Fault diagnosis	66
Chapter 3: Fuel system and lubrication	Specifications	67
	Carburettor	68
	Exhaust system	75
	Lubrication system	75
	Fault diagnosis	78
Chapter 4: Ignition system	Specifications	79
	Flywheel generators	82
	Contact breaker	82
	Ignition timing	84
	Sparking plug	86
	Fault diagnosis	86
Chapter 5: Frame and forks	Specifications	87
	Front forks	87
	Frame	93
	Swinging arm rear fork	93
	Rear suspension units	93
	Speedometer	96
	Cleaning	96
	Fault diagnosis	97
Chapter 6: Wheels, brakes and tyres	Specifications	98
	Front wheel	98
	Brakes	99
	Rear wheel	103
	Brake adjustment	105
	Final drive	106
	Tyres	106
	Fault diagnosis	108
Chapter 7: Electrical system	Specifications	109
	Battery	109
	Fuse	111
	Lights	111
	Wiring	114
	Fault diagnosis	114
	Wiring diagrams	115
Metric conversion tables		118
Index		119

1977 Honda C50 model

1977 Honda C70 model

1977 Honda C90 model

Introduction to the Honda 50, 70 and 90 models

During February 1967 Honda, already well known for their step-through scooter style machine, introduced into the UK their overhead camshaft engine version of the 50 cc motorcycle.

This technical advancement with such a small engine had the result of increasing the performance of the machine to what seemed an incredible level for such a small engine.

Later, in September of the same year, a 90 cc step through model was introduced into the UK for the rider who needed a small reserve of power whilst retaining the same degree of rider protection.

As part of Honda's rationalisation programme the 70 cc step through model was introduced in February 1972, to replace both the 50 cc model and the 90 cc model, being an intermediate size to combine the most desirable features of both machines. Public demand for all three models was such, however, that the 50 cc and 90 cc models were not discontinued and are still currently available. Recently, introduction of 'moped legislation' has caused a decline in the sales of the 50 cc model but further legislation due shortly, may turn the tide of fortune.

General machine specifications

Model	C50	C70	C90
Overall length	1795 mm (70.67 in)	1795 mm (70.67 in)	1830 mm (72.10 in)
Overall width	640 mm (25.19 in)	640 mm (25.19 in)	640 mm (25.19 in)
Overall height	975 mm (38.4 in)	975 mm (38.4 in)	995 mm (39.2 in)
Wheelbase	1185 mm (46.65 in)	1185 mm (46.65 in)	1190 mm (46.89 in)
Ground clearance	130 mm (5.12 in)	130 mm (5.12 in)	130 mm (5.12 in)
Dry weight	68 kg (152 lb)	72 kg (159 lb)	85 kg (187 lb)

Ordering spare parts

When ordering spare parts for any of the Honda C50, C70 or C90 models, it is advisable to deal direct with an official Honda agent, who should be able to supply most items ex-stock. Parts cannot be obtained from Honda (UK) Limited direct; all orders must be routed via an approved agent, even if the parts required are not held in stock.

Always quote the engine and frame numbers in full, particularly if parts are required for any of the earlier models. The frame number is stamped on the left-hand side of the frame, close to the top mounting point of the engine unit. The engine number is stamped on the left-hand crankcase, immediately below the flywheel generator cover.

Use only parts of genuine Honda manufacture. Pattern parts are available, some of which originate from Japan and are packaged to resemble the originals. In many instances these parts will have an adverse effect on performance and/or reliability.

Some of the more expendable parts such as spark plugs, bulbs, tyres, oils and greases etc., can be obtained from accessory shops and motor factors, who have convenient opening hours, charge lower prices and can often be found not far from home. It is also possible to obtain parts on a Mail Order basis from a number of specialists who advertise regularly in the motor cycle magazines.

Frame number location

Engine number location

Routine maintenance

Periodic routine maintenance is a continuous process that commences immediately the machine is used. It must be carried out at specified mileage recordings or on a calendar basis if the machine is not used frequently, whichever is sooner. Maintenance should be regarded as an insurance policy, to help keep the machine in the peak of condition and to ensure long, trouble-free service. It has the additional benefit of giving early warning of any faults that may develop and will act as a safety check, to the obvious advantage of both rider and machine alike.

The various maintenance tasks are described below, under their respective mileage and calendar headings. Accompanying diagrams are provided, where necessary. It should be remembered that the interval between the various maintenance tasks serves only as a guide. As the machine gets older or is used under particularly adverse conditions, it would be advisable to reduce the period between each check.

No special tools are required for the normal routine maintenance tasks. The tools contained in the toolkit supplied with every new machine are limited, but will suffice if the owner wishes to carry out only minor maintenance tasks.

When buying tools, it is worth spending a little more than the minimum to ensure that good quality tools are obtained. Some of the cheaper tools are too soft or flimsy to do an adequate job. It is infuriating to have to stop part way through a job because a spanner has splayed open or broken, and a replacement must be found.

A deep rooted knowledge of engineering principle is by no means necessary before the owner undertakes his or her own maintenance tasks but familiarity with a few of the more commonly used terms and a basic knowledge of how to use tools will help.

The following list of tools will suffice to undertake the routine maintenance tasks described in this Section, but where reference is made to another Chapter for the dismantling procedure, additional tools may be required.

> A tyre pressure gauge.
> A tyre pump
> A 10 mm or 12 mm spark plug spanner
> A set of metric open ended spanners from 6 mm to 17 mm
> A pair of pliers
> Two cross head screwdrivers, size 2 and 3
> A small electrical screwdriver
> A set of feeler gauges
> A 23 mm box or ring spanner
> A 3 mm square socket (Honda tool)
> An adjustable spanner (this tool to be used only as a last resort)

Weekly or every 200 miles (320 km)

1 Check the tyre pressures

The tyre pressure should be 26 psi for the front tyre and 29 psi for the rear tyre measured when the tyres are cold.

Remove the dust cap, flick the valve centre to blow out any dirt or water and push on the pressure gauge. If the pressure is too low, pump up the tyre with the pump or a garage air line to the correct pressure. If the pressure is too high, push the valve centre to release the air until the correct pressure is reached. Replace the dust cap as it is a second seal.

2 Check the engine oil level

The engine oil capacity is 0.8 litres C50 model, 0.7 litre C70 model and 0.9 litre C90 model (1.4, 1.2 and 1.6 pints respectively) contained in a wet sump, and normally SAE 20W/50 but in a cold climate SAE 10W/30 should be used.

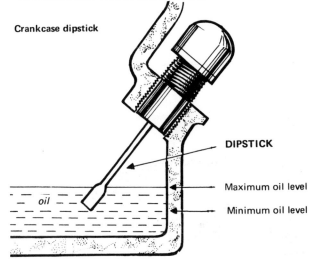

Crankcase dipstick

Check the tyre pressures

Routine maintenance

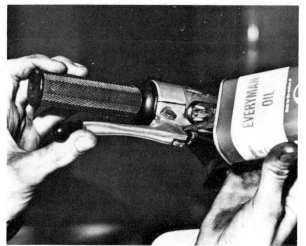

Oiling the front brake lever

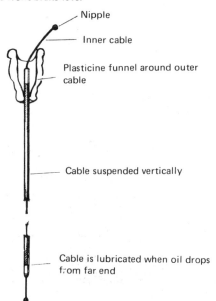

Oiling a control cable

Place the machine on its centre stand on level ground. If the machine has just been run, allow the oil to settle for 5 minutes before checking the level. Remove the plastic filler cap with its integral dipstick. Wipe the oil off the dipstick. Replace the dipstick without screwing it in, remove it and check that the oil level is between the upper and lower limit marks on the dipstick. Add oil if necesssary, to bring the oil to the correct level, and replace the filler cap dipstick after ensuring that the sealing O-ring is in good condition.

3 *Oil and adjust the brake cables and rod*

The standard brake cables should be lubricated with a light machine oil, but if a nylon lined cable has been fitted on no account use oil on it.

Similarly, the cable nipples and pivot points should be oiled including those of the brake rod. Normally, rain and the washing of the machine will provide sufficient lubrication for the nylon and plastic parts. Before the winter sets in each year, it is advisable to remove the cables completely and thoroughly lubricate them as shown in the accompanying sketch, to ensure troublefree riding during the more arduous conditions to be found in winter..

The brakes need adjusting when there is too much movement on the lever or the pedal ie; when the brake lever comes close to the handlebar when the brake is applied or if there is too much movement of the brake pedal. To adjust either brake turn the adjusting nut until the brake just starts to rub when the wheel is spun. Slacken back the adjusting nut until the brake just stops rubbing. Ensure that the adjusting nut cut-outs are seating correctly on the brake operating arms.

4 *Check, adjust and lubricate the final drive chain*

Place the machine on its centre stand on level ground. Check the up and down movement of the chain, midway between the two sprockets. Rotate the back wheel until the up and down movement is at the minimum. This is the 'tight spot' on the chain and the up and down movement here should be between 10 mm (0.40 inch) and 20 mm (0.79 inch).

If the play is greater than 20 mm (0.79 inch) the chain should be adjusted as follows:

Slacken the wheel spindle nuts so that they are finger tight. Make sure that each adjusting nut is turned the same amount, to keep the wheels in line, until the play is reduced to within the limits. Tighten the wheel spindle nuts and recheck the amount of play on the chain.

When the wheels are properly aligned both the adjusters should match the swinging arm markings.

An SAE 90 oil or a **Chain Lubricant** should be spread on the

Oiling the rear brake adjuster

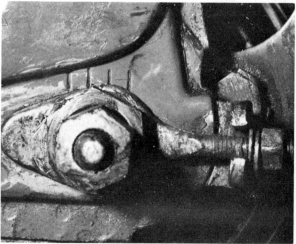

Frame is marked to aid wheel alignment

Routine maintenance

chain for lubrication. The latter is applied from an aerosol pack, to make application easier.

5 *Check the lights and horn*

Check that all the lights are working properly. Renew any defective bulbs and if any lights are dim, clean the connections and earthing points, to restore the lights to their original brightness.

Check that the horn works, again checking the connections if the performance is poor.

6 *Visual inspection*

Give the whole machine a close visual inspection, checking for loose nuts and fittings, frayed control cables or missing parts which may have fallen off or been stolen.

7 *Check the battery*

Remove the side panel, release the battery and slide it out. Remove the top cover, if fitted, and unscrew the three vent plugs on top of the plates. On translucent batteries, the level of the liquid is marked on the outside of the battery. If the liquid level is low, the three separate cells in the battery should be topped up to the correct level with distilled water. Tap water should not be used as the impurities in the water will have an adverse effect on the battery life.

Replace the three plugs and the cover and refit the battery to the machine, finally replacing the side panel.

If any liquid is spilt out of the battery this should be washed off immediately with plenty of water as it will corrode any metal parts and burn the skin if left unattended.

Lubricating the chain

Monthly or every 1000 miles (1600 km)

Check the tyres, brakes, lights and horn as described in the weekly/200 miles service and then carry out the following additional tasks:

1 *Change the engine oil*

As stated previously the engine oil capacity is between 0.7 and 0.9 litres (1.3 and 1.6 pints) of SAE 20W/50 oil, the precise quantity depending on the model.

Place the machine on its centre stand on level ground. Run the engine for a few minutes to warm up the oil so that it will run out easier. Place a container under the engine and remove the drain plug, which is situated on the underside of the engine. When all the oil has drained, replace and tighten the drain plug, ensuring that the sealing washer is in good condition.

Refill the engine with oil of the correct viscosity, checking the level as described in the weekly check.

2 *Check the spark plug*

An NGK type C-7HS or D-6HN spark plug is fitted as standard equipment to the Honda C50, C70 and C90. The recommended gap of the plug is 0.6 mm (0.024 inch) to 0.7 mm (0.028 inch).

Pull off the spark plug cap and unscrew the plug. Clean the electrode to remove any oil or carbon. Check the gap between the electrode with a set of feeler gauges. If the gap needs resetting, bend the outer electrode to bring it closer to the central electrode. Do not try to move the centre electrode as the insulation will break and ruin the plug.

Refit the spark plug and push on the plug cap. Do not overtighten the spark plug as this can cause the thread to strip in the cylinder head. A normal plug spanner has the correct length tommy bar or handle to make overtightening impossible.

A new spark plug should be fitted every 5,000 miles (8,000 km), or if it is damaged or excessively worn.

3 *Check and adjust the throttle cable*

The throttle should have about 10° free play movement. To adjust the amount of free play, slide the rubber sleeve cable, to reveal the adjusting nut. The adjusting nut is turned to provide the correct play and the rubber sleeve, when slid back down the cable, stops the adjusting nut from turning.

4 *Check and adjust the carburettor slow running adjustment*

Any checks or adjustments that are made on the carburettor should be undertaken only when the engine has reached its normal working temperature and not when the engine is cold.

The larger bolt is the drain plug

Refilling the engine with oil

The engine should continue to run slowly when the throttle is closed. If the engine stops every time the throttle is closed, adjustment is necessary. As the machine has an automatic clutch, if the engine runs too fast, the machine will tend to creep forward when it is in gear unless the brakes are applied to stop it.

Slacken the throttle cable to ensure that there is plenty of slack so that cable tension does not give false adjustment on the carburettor.

On the side of the carburettor are two screws, the upper one is the throttle stop screw, the lower the air mixture screw.

To adjust the slow running of the engine, turn the throttle stop screw until the engine is running at approximately 1500 rpm. Turn the air mixture screw until the highest engine speed is obtained. If the engine speed is then too fast, unscrew the throttle stop screw to reduce it, then turn the air mixture screw to find the highest engine speed again. This process is repeated until the engine runs slowly and evenly. Readjust the throttle cable slack to the limit as set out under the previous heading.

5 Check the tyre condition

By law a motorcycle must have a minimum depth of tread of 1 mm (0.04 inch) for at least 75% of the tyre width all the way round the circumference of the tyre. In the interest of safety it is better to renew the tyre long before the legal minimum is reached.

When checking the tyre condition, remove any stones in the tread, check for any bulges, splits or bald spots and renew the tyre if any doubt exists by following the procedure given in Chapter 6, Section 17.

6 Check the clutch adjustment

The clutch plates will wear inside the clutch and the adjustment should be checked periodically to ensure that smooth gearchanging continues.

Clutch adjustment is provided by means of an adjustable screw and locknut located in the centre of the clutch cover. Slacken off the 10 mm locknut and turn the adjusting screw firstly in a clockwise direction, to ensure there is no end pressure on the clutch pushrod.

Turn the adjusting screw anticlockwise until pressure can be felt on the end. Turn back (clockwise) for approximately 1/8th of a turn, and tighten the locknut, making sure the screw does not turn. Clutch adjustment should now be correct.

Six monthly or every 3,000 miles (5,000 km)

Complete all the checks under the weekly and monthly headings and then the following items.

1 Clean the air filter

The air filter is located on top of the main frame tube, immediately behind the steering head, clearly visible when the legshield has been removed.

To clean the air filter, remove the detachable element and tap it lightly to remove accumulated dust. Blow dry from the inside with compressed air, or brush the exterior with a light brush. Remember the element is made from paper. If it is torn or damaged, fit a replacement.

Oil or water will reduce the efficiency of the filter element and may upset the carburettor. Replace any suspect element.

It is advisable to replace the element at less than the recommended 6,000 miles if the machine is used in very dusty conditions. The usual signs of a filter element in need of replacement are reduced performance, misfiring and a tendency for the carburation to run rich.

On no account should the machine be run without the filter element in place because this will have an adverse effect on carburation. Reassembly of the air filter is the reverse of the dismantling procedure.

2 Clean the carburettor and filter

Over a period of time sediment and water will collect in the carburettor. A drain screw on the carburettor enables the float chamber to be flushed out with petrol to remove nearly all of the dirt but Chapter 3 Sections 7 to 10 will describe how the carburettor itself is removed, stripped, cleaned and reassembled, if any trouble still persists.

3 Remove, clean and lubricate the final drive chain

Although the final drive chain is fully enclosed, the oil and grease lubricant on the chain will tend to pick up dust and grit, so every six months it is advisable to remove the chain from the machine for thorough cleaning.

To remove the chain, place the machine on its centre stand on level ground, remove the four bolts and the two chaincase halves and rotate the rear wheel until the spring link is in a convenient position, preferably on the rear wheel sprocket. Use a pair of pliers to remove the spring clip and then remove the side plate and the link plate, thus disconnecting the chain. Connect to one end of the chain a second chain, either an old worn out one or a brand new one which is kept in readiness for fitting to the machine. Pull the first chain off the machine, feeding the second chain on, until the first chain can be disconnected from the second chain. If the second chain is usable, reconnect it, ensuring that the closed end of the spring clip is facing the direction of travel of the chain. Adjust the chain as described in the weekly maintenance section and refit the chaincase.

The chain which has just been removed should be washed thoroughly in petrol or paraffin to remove all the dirt and grease.

To check whether the chain is due for renewal, lay it lengthwise in a straight line and compress it so that all play is taken up. Anchor one end and then pull on the other end to take up the play in the opposite direction. If the chain extends by more than the distance between two adjacent links, it should be renewed in conjunction with the sprockets.

The chain should be lubricated by immersing it in a molten lubricant such as Linklyfe or Chainguard and then hanging it up to drain. This will ensure good penetration of lubricant between the pins and rollers, which is less likely to be thrown off when the chain is in motion.

To refit the chain to the machine, connect it to the second chain, pull the second chain and feed the first chain back onto the machine. Reconnect the chain ensuring the spring clip is correctly fitted as stated before. It is easier to reconnect the chain if the ends are fitted onto the rear wheel sprocket whilst the connecting link is inserted. Adjust the chain, using the weekly maintenance procedure, and refit the chaincase.

4 Check and adjust the valve tappet clearances

The valve tappet clearance for both the inlet and exhaust valves is 0.05 mm (0.002 inch) when the engine is cold.

A small amount of dismantling is required before the tappet clearance can be checked.

Ensure that the machine is on its centre stand, standing on level ground. Remove the flywheel inspection cover on the left-hand side of the engine. Remove also the two tappet covers to reveal the adjusters.

To check the tappet clearances, turn the flywheel until the line marked with a 'T' is aligned with the mark on the flywheel cover. The piston will now be at top dead centre on either the compression or exhaust stroke. Checking the tappet clearances must be made on the compression stroke when both rocker arms are free to rock, so a complete turn of the flywheel is required if the piston is on the exhaust stroke. It will probably be found that when turning the flywheel, the 'T' mark tends to move on every other revolution when the piston is under compression. This is the position required for checking the tappet clearances and to avoid the 'T' mark moving, removing the spark plug and its cover will relieve the pressure in the cylinder.

A 0.05 mm (0.002 inch) feeler gauge should just pass between the rocker arm and the valve stem. If adjustment is necessary, slacken the locking nut and turn the adjusting screw until the feeler gauge will just pass through the gap. Hold the adjusting screw securely and retighten the locknut. Check the gap again

Align the 'T' mark with the mark on the cover

Use a feeler gauge to check the gap

Align the 'F' mark with the mark on the cover

to ensure that it is still correct. This applies to both valves as the clearance is identical.

Refit the tappet covers, checking the condition of the O-rings. Refit the spark plug, the plug cover and the flywheel cover, unless the next task of checking the ignition timing is about to be carried out.

5 Check and adjust the ignition timing

C50 and C70 models only: The ignition timing is determined by when the contact breaker points open. The flywheel operates the contact breaker and the heel of the contact arm will wear, altering the ignition timing. The flywheel inspection cover should be removed so that the contact breaker can be viewed through one of the apertures in the flywheel. When the line marked 'F' on the flywheel lines up with the mark on the crankcase shaft, the contact breaker should just start to open. If adjustment is necessary, the fixed contact can be moved by slackening the clamping screw and using a screwdriver in the slot provided. Retighten the clamping screw and check the adjustment again, to ensure that it has not altered.

When the ignition timing is correct, rotate the flywheel to determine the position at which the contact breaker points are fully open. When fully open the contact breaker gap should be between 0.3 mm and 0.4 mm (0.012 and 0.016 in).

If the gap is too small, the contact breaker points need renewing, as described in Chapter 4 Section 6.

Refit the flywheel cover and the spark plug and plug cover, if these have been removed.

C90 model only: The ignition timing is determined by when the contact breaker points open. The camshaft operates the contact breaker and the heel of the contact arm will wear, altering the ignition timing. The flywheel inspection cover and the contact breaker cover must be removed. Before checking the ignition timing, the contact breaker gap should be checked. Rotate the engine until the contact breaker is in its fully open position. Check the gap to see if it is between 0.3 and 0.4 mm (0.012 and 0.016 inch). To adjust the gap, slacken the two screws that hold the contact breaker assembly, and using a small screwdriver in the slot provided ease the assembly to the correct position. Tighten the screws and recheck the gap to ensure that the assembly has not moved.

The ignition timing is correct when the contact breaker points are about to separate when the 'F' line scribed on the flywheel rotor coincides exactly with the mark on the cover. The backplate holding the complete contact breaker assembly is slotted, to permit a limited range of adjustment. If the two crosshead retaining screws are slackened a little, the plate can be turned

Check the return spring condition

Greasing the suspension pivots

Greasing the leading link fork pivots

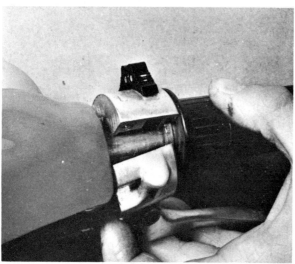
Remove the switch assembly to release the twistgrip

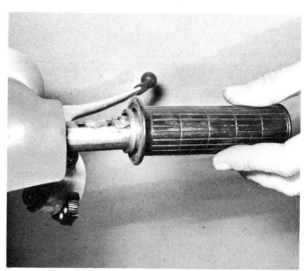
Slide the twistgrip off for greasing

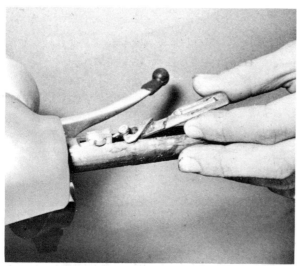
Remove the sliding block to oil the cable

Oil the felt wick

Routine maintenance

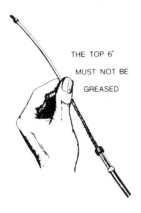

Greasing the speedometer cable

until the points commence to separate, and then locked in this position by tightening the screws.

After checking the timing, rotate the engine and check again before replacing the covers. The accuracy of the ignition timing is critical in terms of both engine performance and petrol consumption. Even a small error in setting can have a noticeable effect on both.

6 *Check the centre and prop stand springs*

Check the condition of the centre and prop stand springs and renew them if they are worn or heavily corroded. If the stand drops when the machine is moving, it may catch in some obstacle in the road and unseat the rider. Grease the springs and the stand pivot points.

7 *Grease the speedometer cable*

Once the headlamp lens has been removed the speedometer cable can be unscrewed from the speedometer head and pulled clear. The inner cable can then be pulled out. Clean off the old grease by washing in petrol or paraffin. Spread new grease along the length of the cable except for the top 15 cm (6 inch) and feed the cable back into the outer casing. Reconnect the cable to the speedometer head.

If the top of the cable is greased, the grease will work its way into the speedometer head and stop it functioning, thus necessitating a replacement as the speedometer head cannot be stripped for cleaning.

8 *Grease the leading link forks*

Apply a grease gun to the nipples on the leading links and the fork blades. Pump grease into the bearings until clean grease emerges from the joints. Wipe away any excess grease as this will collect the dirt and make the machine look unsightly.

9 *Grease the throttle twistgrip and oil the throttle cable*

Remove the two screws and the indicator switch from the handlebar. This releases the twisting sleeve which slides off the handlebar with an anti-clockwise twist. The sliding block can now be lifted out of the handlebar and the inner cable unhooked. The outer cable stop can then be removed from the handlebar and unhooked from the cable.

Yearly or every 5,000 miles (8,000 km)

Complete all the checks listed under the weekly, monthly and six monthly headings, then complete the following additional tasks:

1 *Lubricate the felt wick of the contact breaker cam*

When the ignition timing is checked, the felt wick of the contact breaker cam can be seen. A few drops of light machine oil, should be put on the wick to reduce wear on the heel of the contact arm. Do not over oil. If oil finds its way on to the contact breaker points it will act as an insulator and prevent electrical contact from being made.

2 *Check the condition of the sprockets*

When the final drive chain is cleaned and checked, ensure that the sprockets are not badly worn, before replacing the chain. If the sprocket teeth are badly worn, they will probably have a hooked appearance and should be renewed as described in Chapter 1 Sections 16 and 47 and Chapter 6 Section 12.

3 *Adjust and lubricate the steering head bearings*

Dismantling and reassembly of the steering head is a task that should be undertaken only if a good understanding of the problems involved is realised. Chapter 5 Sections 2 and 4, fully describe the necessary procedures.

4 *Examine and lubricate the wheel bearings*

Dismantling and reassembly of the wheel bearings is also a task to be undertaking only if an understanding of the problems involved is realised. Chapter 6 fully describes the necessary procedure.

5 *Examine both front and rear brake assemblies*

The brake assemblies should be cleaned to remove any dust and checked to ensure that the brake linings are not wearing too thin. This task is fully described in Chapter 6.

Recommended lubricants

Component	Type of lubricant
Engine	
Normal temperature	Multi-grade 20W/50 engine oil
Below freezing point	Multi-grade 10W/30 engine oil
Final drive chain	Graphited grease or aerosol type chain lubricant
All greasing points	Multi-purpose high melting point lithium-based grease
Oil points	Light oil

The engine oil should be changed every 1,000 miles. In winter, or when the machine is used for short journeys only, the oil must be changed every 300 miles.

Working conditions and tools

When a major overhaul is contemplated, it is important that a clean, well-lit working space is available, equipped with a workbench and vice, and with space for laying out or storing the dismantled assemblies in an orderly manner where they are unlikely to be disturbed. The use of a good workshop will give the satisfaction of work done in comfort and without haste, where there is little chance of the machine being dismantled and reassembled in anything other than clean surroundings. Unfortunately, these ideal working conditions are not always practicable and under these latter circumstances when improvisation is called for, extra care and time will be needed.

The other essential requirement is a comprehensive set of good quality tools. Quality is of prime importance since cheap tools will prove expensive in the long run if they slip or break and damage the components to which they are applied. A good quality tool will last a long time, and more than justify the cost. The basis of any tool kit is a set of open-ended spanners, which can be used on almost any part of the machine to which there is reasonable access. A set of ring spanners makes a useful addition, since they can be used on nuts that are very tight or where access is restricted. Where the cost has to be kept within reasonable bounds, a compromise can be effected with a set of combination spanners - open-ended at one end and having a ring of the same size on the other end. Socket spanners may also be considered a good investment, a basic 3/8 or ½ inch drive kit comprising a ratchet handle and a small number of socket heads, if money is limited. Additional sockets can be purchased, as and when they are required. Provided they are slim in profile, sockets will reach nuts or bolts that are deeply recessed. When purchasing spanners of any kind, make sure the correct size standard is purchased. Almost all machines manufactured outside the UK and USA have metric nuts and bolts, whilst those produced in Britain have BSF or BSW sizes. The standard used in the USA is AF, which is also found on some of the later British machines. Other tools that should be included in the kit are a range of crosshead screwdrivers, a pair of pliers and a hammer.

When considering the purchase of tools, it should be remembered that by carrying out the work oneself, a large proportion of the normal repair cost, made up by labour charges, will be saved. The economy made on even a minor overhaul will go a long way towards the improvement of a tool kit.

In addition to the basic tool kit, certain additional tools can prove invaluable when they are close to hand to help speed up a multitude of repetitive jobs. For example, an impact screwdriver will ease the removal of screws that have been tightened by a similar tool, during assembly, without risk of damaging the screw heads. And, of course, it can be used again to retighten the screws, to ensure an oil or airtight seal results. Circlip pliers have their uses too, since gear pinions, shafts and similar components are frequently retained by circlips that are not too easily displaced by a screwdriver. There are two types of circlip pliers, one for internal and one for external circlips. They may also have straight or right-angled jaws.

One of the most useful of all tools is the torque wrench, a form of spanner that can be adjusted to slip when a measured amount of force is applied to any bolt or nut. Torque wrench settings are given in almost every modern workshop or service manual, where the extent to which a complex component, such as a cylinder head, can be tightened without fear of distortion or leakage. The tightening of bearing caps is yet another example. Overtightening will stretch or even break bolts, necessitating extra work to extract the broken portions.

As may be expected, the more sophisticated the machine, the greater is the number of tools likely to be required if it is to be kept in first class condition by the home mechanic. Unfortunately there are certain jobs which cannot be accomplished successfully without the correct equipment and although there is invariably a specialist who will undertake the work for a fee, the home mechanic will have to dig more deeply in his pockets for the purchase of similar equipment if he does not wish to employ the service of others. Here a word of caution is necessary, since some of these jobs are best left to the expert. Although an electrical multimeter of the Avo type will prove helpful in tracing electrical faults, in inexperienced hands it may irrevocably damage some of the electrical components if a test current is passed through them in the wrong direction. This can apply to the sychronisation of twin or multiple carburettors too, where a certain amount of expertise is needed when setting them up with vacuum gauges. These are, however, exceptions. Some instruments, such as a strobe lamp, are virtually essential when checking the timing of a machine powered by CDI ignition system. In short, do not purchase any of these special items unless you have the experience to use them correctly.

Although this manual shows how components can be removed and replaced without the use of special service tools (unless absolutely essential), it is worthwhile giving consideration to the purchase of the more commonly used tools if the machine is regarded as a long term purchase. Whilst the alternative methods suggested will remove and replace parts without risk of damage, the use of the special tools recommended and sold by the manufacturer will invariably save time.

Chapter 1 Engine and gearbox

Contents

General description ... 1	Kickstart assembly: examination ... 35
Operations with engine in frame ... 2	Primary drive gears: examination ... 36
Operations with engine removed ... 3	Gear selector drum and gear cluster: examination ... 37
Method of engine/gearbox removal ... 4	Engine reassembly: general ... 38
Removing the engine/gearbox unit ... 5	Engine reassembly: fitting the bearings and oil seals to the crankshaft and clutch cover ... 39
Dismantling the engine and gearbox: general ... 6	
Engine and gearbox dismantling: removal of generator ... 7	Engine reassembly: refitting the oil pump ... 40
Cylinder head and cylinder: removal ... 8	Engine reassembly: refitting the camchain tensioner pulley and oil pump drive ... 41
Piston and piston rings: removal ... 9	
Camshaft and rocker arms: removal ... 10	Engine reassembly: refitting neutral indicator switch ... 42
Valves and valve guides: removal ... 11	Engine reassembly: replacing the gear selector drum and gear cluster ... 43
Oil filters: removal ... 12	
Clutch and primary drive: removal ... 13	Engine reassembly: replacing the kickstart shaft assembly ... 44
Gearchange mechanism: removal ... 14	Engine reassembly: replacing the crankshaft assembly ... 45
Kickstart return spring: removal ... 15	Engine reassembly: rejoining the crankcases ... 46
Final drive sprocket: removal ... 16	Engine reassembly: refitting the final drive sprocket ... 47
Crankcases: separating ... 17	Engine reassembly: replacing and tensioning the kickstart return spring ... 48
Crankshaft assembly: removal ... 18	
Kickstart shaft assembly: removal ... 19	Engine reassembly: refitting the gearchange mechanism ... 49
Gear selector drum and gear cluster: removal ... 20	Engine reassembly: reassembling the clutch and primary drive ... 50
Neutral indicator switch: removal ... 21	
Camchain tensioner pulley and oil pump drive: removal ... 22	Engine reassembly: replacing the oil filters ... 51
Oil pump: removal ... 23	Engine reassembly: replacing the valves and valve guides ... 52
Oil seals: removal ... 24	Engine reassembly: replacing the camshaft and rocker arms ... 53
Crankshaft and gearbox main bearings: removal ... 25	
Examination and renovation: general ... 26	Engine reassembly: refitting the piston and cylinder barrel ... 54
Big-end and main bearings: examination and renovation ... 27	
Cylinder barrel: examination and renovation ... 28	Engine reassembly: replacing the cylinder head and timing the valves ... 55
Piston and rings: examination and renovation ... 29	
Valves, valve seats and valve guides: examination and renovation ... 30	Engine reassembly: adjusting the tappets ... 56
	Engine reassembly: replacing the flywheel generator ... 57
Cylinder head: decarbonisation and examination ... 31	Refitting the engine/gearbox unit in the frame ... 58
Camshaft, rockers and rocker shafts: examination ... 32	Starting and running the rebuilt engine ... 59
Camshaft chain tension and sprockets: examination ... 33	Fault diagnosis: engine ... 60
Gearchange mechanism: examination ... 34	Fault diagnosis: gearbox ... 61

Specifications

Engine

Type	Single cylinder overhead camshaft, chain operated
Cylinder head	Aluminium alloy
Cylinder barrel	Cast iron
Bore	39 mm (C50)
	47 mm (C70)
	50 mm (C90)
Stroke	41.4 mm (C50 and C70)
	45.6 mm (C90)
Capacity	49 cc (C50)
	72 cc (C70)
	89.6 cc (C90)
Bhp	4.8 @ 10,000 rpm (C50)
	6.2 @ 9,000 rpm (C70)
	7.5 @ 9,500 rpm (C90)

Compression ratio	8.1 : 1 (C70) 8.2 : 1 (C90) 8.8 : 1 (C50)

Crankshaft
Crankpin outside diameter	23.098 - 23.112 mm (0.9099 - 0.9105 inch)

Connecting rod
Big-end end float	0.1 - 0.35 mm (0.004 - 0.014 inch)
Small end and gudgeon pin clearance	0.025 - 0.050 mm (0.001 - 0.002 inch)
Small end bore diameter	13.016 - 13.043 mm (0.5124 - 0.5135 inch) (50 and 70 cc models) 14.012 - 14.028 mm (0.5517 - 0.5523 inch) (90 cc model)

Piston
Maximum diameter at base of skirt	38.98 - 39.00 mm (1.5392 - 1.5400 inch) (50 cc model) 46.98 - 47.00 mm (1.8492 - 1.8500 inch) (70 cc model) 49.97 - 49.99 mm (1.9673 - 1.9681 inch) (90 cc model)
Piston to cylinder clearance (minimum)	0.01 mm (0.0004 inch) Replace if over 0.1 mm (0.004 inch)
Oversize pistons available	+0.25 mm, +0.50 mm, +0.75 mm and +1.00 mm
Piston gudgeon pin clearance	0.002 - 0.014 mm (0.00008 - 0.00055 inch)

Piston rings
Compression (two top rings)	Top ring chrome, second ring tapered
Oil control ring	Third ring at top of skirt

Gudgeon pin
Diameter	13.002 - 13.008 mm (0.5121 - 0.5123 inch)

Valves
Tappet clearance, inlet and exhaust	0.05 mm (0.002 inch) set with engine cold
Seat angle	45°
Inlet, overall length	64.5 mm (2.540 inch) (70 cc model) 66 mm (2.600 inch) (50 cc model) 67.3 mm (2.648 inch) (90 cc model)
Outside diameter of stem	5.5 mm (0.217 inch) (50 and 70 cc models) 5.45 mm (0.215 inch) (90 cc model)
Exhaust, overall length	63.9 mm (2.483 inch) (70 cc model) 65.3 mm (2.573 inch) (50 cc model) 67.3 mm (2.597 inch) (90 cc model)
Ouside diameter of stem	5.5 mm (0.217 inch) (50 and 70 cc models) 5.43 mm (0.214 inch) (90 cc model)
Stem and guide clearance (inlet)	0.010 - 0.030 mm (0.0004 - 0.0012 inch)
Stem and guide clearance (exhaust)	0.030 - 0.050 mm (0.0012 - 0.0020 inch)
Spring (outer) free length	28.1 mm (1.110 inch) (50 and 70 cc models) 31.8 mm (1.253 inch) (90 cc model)
Spring (inner) free length	25.5 mm (1.01 inch) (50 and 70 cc models) 26.5 mm (1.044 inch) (90 cc model)

Capacities
Engine and gearbox (in unit)	50 cc model	0.8 litres (1.4 Imp pints) (1.7 US pints)
	70 cc model	0.7 litres (1.2 Imp pints) (1.5 US pints)
	90 cc model	0.9 litres (1.58 Imp pints) (1.90 US pints)

Torque wrench settings
Cylinder head nuts	6.5 - 8.7 ft lb
Cylinder head left side cover	5.8 - 8.7 ft lb
Cylinder head right-hand side cover	5.1 - 6.5 lb ft
Carburettor mounting studs	60 in lb
Flywheel nut	23.9 - 27.5 ft lb
Clutch nut	27.5 - 32.5 lb ft

Chapter 1: Engine and gearbox

1 General description

The Honda C50, C70 and C90 models are fitted with an overhead camshaft engine in which the valve mechanism is chain driven. The camshaft is located within the aluminium alloy cylinder head; with this arrangement it is necessary to disturb the valve timing when the cylinder head is removed.

All engine/gear units are of aluminium alloy construction, with a cast iron cylinder barrel. The flywheel generator is mounted on the left-hand side of the engine unit; the clutch assembly is located on the right-hand side of the engine, behind a domed aluminium alloy cover. Convention is defied by installing the engine/gear unit in a near-horizontal position, so that the cylinder barrel is almost parallel to the ground. The exhaust system is carried on the right-hand side of the machine, of the down-swept pattern. All models are fitted with a conventional kickstart.

A trochoid oil pump is included in the general specification of the engine, to provide a pressure oil feed in addition to lubrication by splash. There are also two filters in the lubrication system, a gauze screen in the crankcase and a centrifugal filter within the clutch centre.

The engine is built in unit with the gearbox. This means that when the engine is dismantled, the gearbox has to be dismantled too, and vice-versa.

2 Operations with engine/gearbox in frame

It is not necessary to remove the engine unit from the frame unless the crankshaft assembly and/or the gearbox bearings require attention. Most operations can be accomplished with the engine in place, such as:
1. Removal and replacement of cylinder head.
2. Removal and replacement of cylinder barrel and piston.
3. Removal and replacement of flywheel magnetic generator.
4. Removal and replacement of clutch assembly.
5. Removal and replacement of timing pinions and kickstart assembly.

When several operations need to be undertaken simultaneously, it will probably be advantageous to remove the complete engine unit from the frame, an operation that should take approximately twenty minutes. This will give the advantage of better access and more working space.

3 Operations with engine/gearbox removed

1. Removal and replacement of the main bearings.
2. Removal and replacement of the crankshaft assembly.
3. Removal and replacement of the gear cluster, selectors and gearbox main bearings.

4 Method of engine/gearbox removal

As described previously, the engine and gearbox are built in unit and it is necessary to remove the complete unit in order to gain access to either component. Separation is accomplished after the engine unit has been removed and refitting cannot take place until the crankcase has been reassembled.

5 Removing the engine/gearbox unit

1. Place the machine on the centre stand and make sure it is standing firmly, on level ground.
2. Remove the side panels to obtain the tool kit and reveal the battery. Unscrew the fuse holder and remove the fuse, to isolate the battery.
3. Remove the domed nut and the air cleaner lid and withdraw the element.

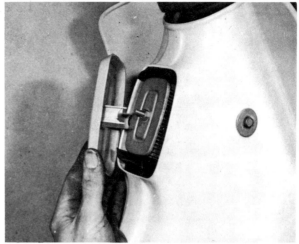
5.3 Remove the air cleaner lid

5.4a Slacken the nuts and remove the clamping band ...

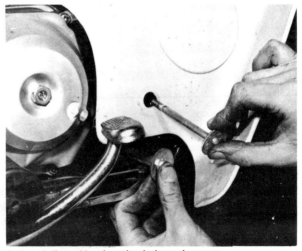
5.4b ... followed by the other bolts and spacers

5.6 Release the exhaust pipe from the cylinder

5.8 Remove the gearchange lever

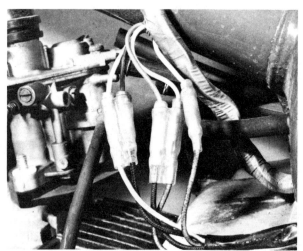
5.10 Disconnect the wires at the snap connectors

5.11 Remove the spark plug cap and HT lead clip

5.12 Disconnect the brake and stoplight springs

4 Slacken the nuts holding the rear of the plastic legshields and remove the clamping band, if fitted. Remove the four bolts holding the legshields and pull their spacers clear. The legshields will now lift clear, allowing easy access to the engine.
5 Remove the crankcase drain plug and drain the oil into a suitable container.
6 Remove the two nuts holding the exhaust pipe to the cylinder head and unscrew the swinging arm pivot nut to release the silencer. The exhaust system will then pull clear, pulling the two collets out of the cylinder head. Remove the copper/asbestos joint ring from the exhaust port.
7 Ensure that the petrol is switched off and remove the nuts clamping the carburettor to the cylinder head. The carburettor can be left in place when the engine is dropped out of the frame.
8 Remove the gearchange lever clamp bolt and slide the lever off its shaft.
9 Remove the three screws and the flywheel generator cover or two screws and rear cover on the C90 model, to reveal the final drive sprocket. Rotate the rear wheel until the chain spring link is in a removable position. Disconnect the chain at the spring link and pull the chain clear of the engine sprocket. If the chain is reconnected, it will ease finding both ends of the chain when reassembling. It may be necessary to remove two bolts and the top half of the chainguard to obtain greater access to the chain.

Chapter 1: Engine and gearbox

10 Pull the snap connectors apart to disconnect the engine wires. All of them are colour coded for easy reconnection.
11 Pull off the spark plug cap and remove the screw to release the HT lead.
12 Unhook the brake pedal and stop lamp switch springs.
13 Remove the footrest assembly, which is attached to the crankcase by four 14 mm bolts and spring washers. It is possible to remove the engine/gear unit with the footrests in place, if it is desired to use them as a convenient carrying handle.
14 Remove the top engine bolt and allow the engine to pivot down. Ensure that the carburettor has slid off its studs and is clear of the engine. Remove the bottom engine bolt and pull the engine clear of the machine. Note that the C90 model has an inlet tube bolted to the cylinder head which can either be removed or left attached to the carburettor.

6 Dismantling the engine and gearbox: general

Before commencing work on the engine unit, the external surface should be cleaned thoroughly. A motor cycle engine has very little protection from road grit and other foreign matter, which will find its way into the dismantled engine if this simple precaution is not observed. One of the proprietary engine cleaning compounds such as 'Gunk' or 'Jizer' can be used to good effect particularly if the compound is allowed to work into the film of grease and oil before it is washed away. When washing down, make sure that water cannot enter the carburettor or the electrical system, particularly if these parts have been exposed.

Never use undue force to remove any stubborn part, unless mention is made of this requirement. There is invariable good reason why a part is difficult to remove, often because the dismantling operation has been tackled in the wrong sequence.

Dismantling will be made easier if a simple engine stand is constructed that will correspond with the engine mounting points. This arrangement will permit the complete unit to be clamped rigidly to the work bench, leaving both hands free.

7 Dismantling the engine and gearbox: removal of generator

Engine in the frame

As stated in Section 2 of this Chapter, it is possible to remove the generator whilst the engine is still in the frame. Only paragraphs 1 to 5, 8 and 10 of Section 5 need to be completed before proceeding with the following dismantling procedure;

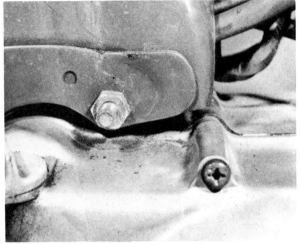
5.14a Remove the top engine bolt ...

5.14b ... followed by the bottom engine bolt

7.2 Remove the rotor nut

7.3 Method for removing the rotor when special service tool is not available

Chapter 1: Engine and gearbox

7.4 Release the wire on the neutral switch

7.6 Remove the stator plate and take care not to lose the Woodruff key

8.2 Remove the circular cover

Engine removed from the frame

If the whole of Section 5 has been completed, continue with the following dismantling procedure:

C50 and C70 models only

1 Remove the three screws and the generator cover.
2 Hold the generator securely and remove the central nut and washer.
3 Use a Honda extractor tool to remove the rotor as it is a keyed taper fit. If the extractor is not available, it may be possible to use the method shown in the accompanying photograph. Wrap some emery cloth round the rotor and clamp onto the rotor two hose clips joined end to end. A two legged sprocket puller hooked onto the hose clips can then be used to pull off the rotor.
4 Disconnect the green/red striped wire from the neutral indicator switch.
5 Remove the two countersunk crosshead screws holding the stator plate to the crankcase and remove the stator plate complete with wires. If scribe marks are made across the stator plate and its housing, this will aid reassembly and may obviate the need to retime the ignition.
6 Remove the Woodruff key from the crankshaft, and collect the two small O-rings that seal the stator plate screws.

C90 model only

7 Remove the two screws and the final drive sprocket cover. This allows the eight screws and the generator cover to be removed.
8 Disconnect the green/red striped wire from the neutral indicator contact and remove the complete stator coil assembly from the left-hand crankcase by unscrewing the crosshead screws.
9 Hold the rotor stationary and remove the centre retaining bolt and washer.
10 Use a Honda extractor tool to remove the rotor. If the correct service tool extractor is not available, use a sprocket puller on a bolt screwed part way into the end of the crankshaft.

8 Cylinder head and cylinder: removal

Engine in the frame

As stated in Section 2 of this Chapter, it is possible to remove the cylinder head and the cylinder barrel whilst the engine is still in the frame. Only paragraphs 1 to 7 of Section 5 need to be completed before proceeding with the following dismantling procedure:

Engine removed from the frame

If the whole of Section 5 has been completed continue with the following dismantling procedure:

C50 and C70 models only

1 Remove the spark plug cap and unscrew the spark plug.
2 Remove the long bolt which passes through the centre of the camshaft and pull off the circular side cover on the left-hand side of the cylinder head.
3 Rotate the engine until the 'O' mark on the camshaft sprocket lines up with the notch on the cylinder head. This ensures that the engine is at top dead centre (TDC) on the compression stroke.
4 Remove the sealing plug and the camchain tensioner spring from the underside of the engine.
5 Remove the three bolts and the camshaft sprocket from the end of the camshaft.
6 Remove the four nuts and washers from the top of the engine, noting the position of the domed nuts and sealing washers. The top engine cover will now lift clear.
7 Remove the single bolt on the left-hand side of the engine and slide the cylinder head up the holding down studs, allowing the camshaft sprocket to drop clear of the cylinder head.
8 Remove the sprocket from the chain and remove the cylinder head gasket and its associated 'O' rings.
9 Remove the bearing bolt for the camchain guide roller and pull the roller clear.
10 Remove the single bolt on the left-hand side of the engine and

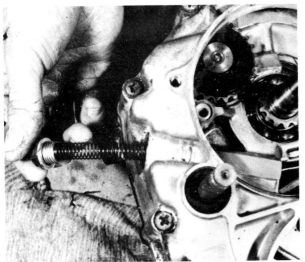
8.4 Remove the tensioner spring

8.5 Remove the three camshaft bolts

8.6 Remove the nuts and top cover

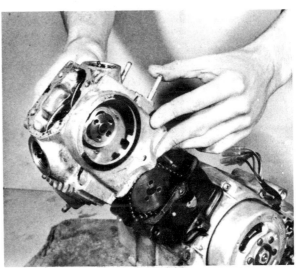
8.7 Slide the cylinder head up the studs

8.8 Remove the camshaft sprocket

8.10 Pad the crankcase mouth and slide off the barrel

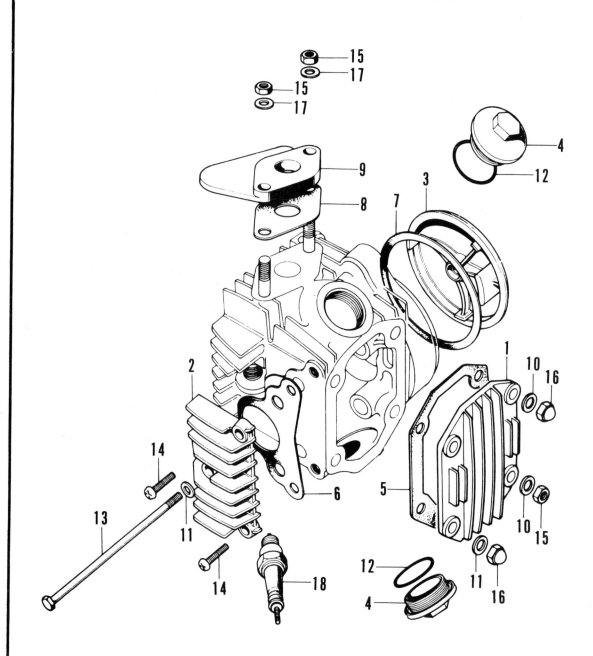

Fig. 1.1. Cylinder head covers - C70 model

1 Top cover
2 Right-hand cover
3 Left-hand cover
4 Tappet cover (2 off)
5 Top cover gasket
6 Right-hand cover gasket
7 Left-hand cover gasket
8 Insulator block gasket
9 Insulator block
10 Washer (3 off)
11 Washer (2 off)
12 O-ring (2 off)
13 Bolt
14 Screw (2 off)
15 Nut (3 off)
16 Dome nut (3 off)
17 Washer (2 off)
18 Spark plug

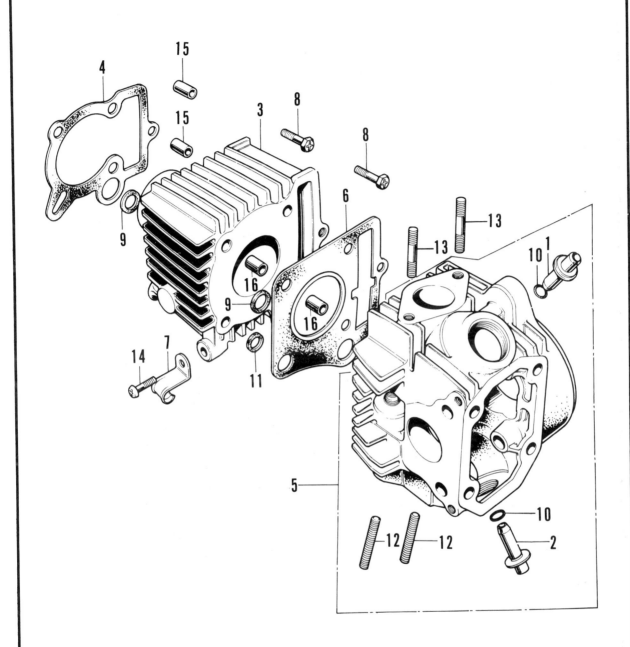

Fig. 1.2. Cylinder head and barrel - C70 model

1 Inlet valve guide
2 Exhaust valve guide
3 Cylinder barrel
4 Cylinder base gasket
5 Cylinder head assembly
6 Cylinder head gasket
7 Clip
8 Bolt (2 off)
9 Rubber seal (2 off)
10 O-ring (2 off)
11 Rubber seal
12 Stud (2 off)
13 Stud (2 off)
14 Screw
15 Dowel (2 off)
16 Dowel (2 off)

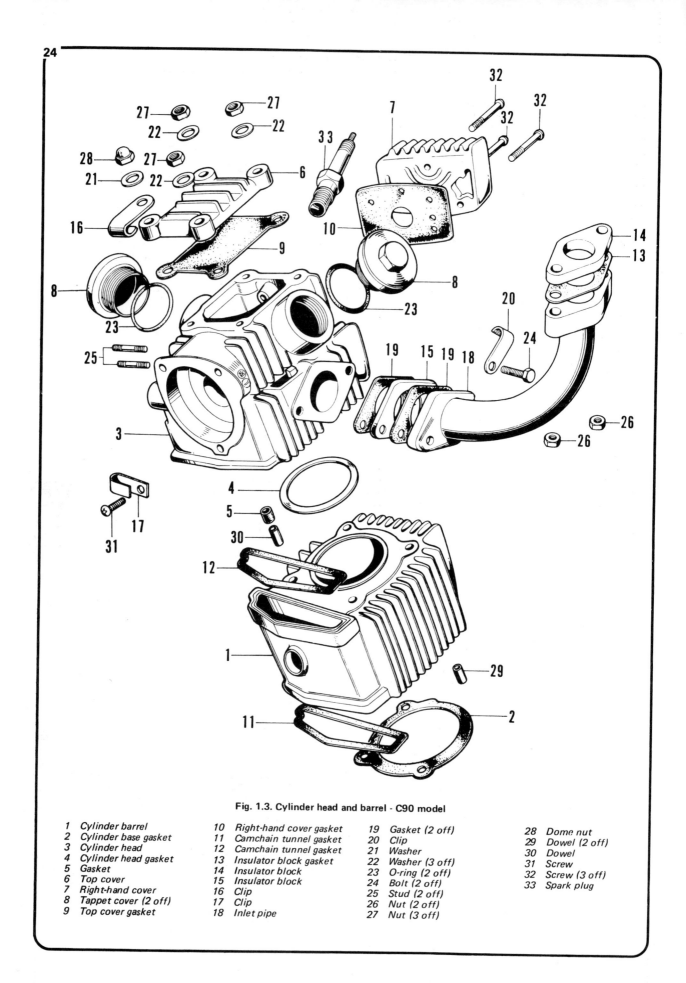

Fig. 1.3. Cylinder head and barrel - C90 model

1 Cylinder barrel	10 Right-hand cover gasket	19 Gasket (2 off)	28 Dome nut
2 Cylinder base gasket	11 Camchain tunnel gasket	20 Clip	29 Dowel (2 off)
3 Cylinder head	12 Camchain tunnel gasket	21 Washer	30 Dowel
4 Cylinder head gasket	13 Insulator block gasket	22 Washer (3 off)	31 Screw
5 Gasket	14 Insulator block	23 O-ring (2 off)	32 Screw (3 off)
6 Top cover	15 Insulator block	24 Bolt (2 off)	33 Spark plug
7 Right-hand cover	16 Clip	25 Stud (2 off)	
8 Tappet cover (2 off)	17 Clip	26 Nut (2 off)	
9 Top cover gasket	18 Inlet pipe	27 Nut (3 off)	

Chapter 1: Engine and gearbox

slide the cylinder barrel up the holding down studs sufficiently to enable the crankcase mouth to be padded with a clean rag to stop any broken pieces or dirt falling inside the engine which would necessitate further engine dismantling to remove them. Slide the cylinder barrel further up the studs and support the piston as it falls clear of the barrel. Remove the barrel completely followed by the cylinder base gasket and 'O' ring.

C90 model only

The contact breaker assembly of this model is located within the cylinder head casting, where it is driven from an extension of the overhead camshaft. In consequence, a special dismantling procedure is necessary, as follows:

11 Remove the circular contact breaker cover on the cylinder head, held by two crosshead screws.
12 Disconnect the lead wire to the contact breaker assembly and remove the contact breaker assembly complete with back plate. It is retained in position by two crosshead screws, which should be removed. If the exact position of the back plate is marked with a scribe line in relation to its housing, this will aid reassembly and possibly obviate the need to retime the ignition.
13 Remove the automatic advance unit by withdrawing the hexagon head bolt from the centre of the camshaft. Remove also the dowel pin, which is used to ensure the assembly is replaced in the correct position.
14 Detach the contact breaker outer casting and gasket, which is held to the cylinder head casting by three crosshead screws.
15 Remove the spark plug cap and unscrew the spark plug.
16 Rotate the engine until the 'O' mark on the camshaft sprocket lines up with the notch on the cylinder head. This ensures that the engine is at top dead centre (TDC) on the compression stroke.
17 Remove the sealing plug and the camchain tensioner spring from the underside of the engine.
18 Remove the two bolts and pull the camshaft clear of the head leaving the chain and sprocket within the cylinder head.
19 Remove the four nuts and washers from the top of the engine, noting the positions of the domed nuts and sealing washers. The top engine cover will now lift clear.
20 Slide the cylinder head up the holding down studs, allowing the camshaft sprocket to drop clear of the cylinder head.
21 Remove the sprocket from the chain and remove the cylinder head gasket, the oil feed seal and the camchain tunnel seal.
22 Remove the bearing bolt for the camchain guide roller and pull the roller clear.
23 Slide the cylinder barrel up the holding down studs sufficiently to enable the crankcase mouth to be padded with a clean rag, to stop any broken pieces or dirt falling inside the engine which would necessitate further engine dismantling to remove them. Slide the cylinder barrel further up the studs and support the piston as it falls clear of the barrel. Remove the barrel completely followed by the cylinder base gasket and the camchain tunnel seal.
24 If the flywheel generator has been removed, the camchain can then be detached.

9 Piston and piston rings: removal

1 The gudgeon pin is of the fully floating type, retained by two wire circlips in the piston bosses. After the circlips have been removed, using pointed nose pliers, the pin can be tapped lightly from the piston.
2 Note the piston is marked with an arrow and must be positioned so that the arrow points downwards. If the piston is oversize, the amount will be stamped on the piston crown.
3 Remove the piston rings by expanding them gently, using extreme care because they are very brittle. If they prove difficult to remove, slide strips of tin behind them, to help ease them from their grooves. The top piston ring is of the chrome type and should have the mark 'top' on the uppermost face. The second ring is tapered and should have the 'top' mark in a similar position. A slotted oil scraper ring is fitted in the lower groove,

9.2 The piston is marked with an arrow

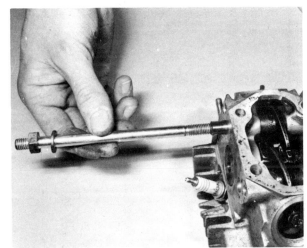
10.2a Use a bolt to withdraw the rocker pins ...

10.2b ... to release the rocker arms

which can be located with either face uppermost.

10 Camshaft and rocker arms: removal

1 The camshaft on the C50 and C70 models slides out of the cylinder head once the cam lobes have been lined up with their respective cutouts. The C90 model camshaft will have been removed in the cylinder head removal sequence.
2 Remove the two or three (C90 model) screws and the right-hand (finned) cover. Pull out the rocker pins by using a standard bolt screwed into the extraction thread. The engine mounting bolt is of the correct thread size. The rocker arms will now pull clear.

11 Valves and valve guides: removal

1 Remove the two tappet covers from the cylinder head.
2 Use a small size valve spring compressor to compress the springs and release the two half collets.
3 Release the valve spring compressor and remove the top spring register, the inner and outer valve springs, and on the exhaust valve only, the bottom spring register, the oil seal cover and the oil seal.
4 The valve will slide out of its guide. The other valve can now be treated in exactly the same way.
5 If it is necessary to remove the valve guides, they can be tapped out with a hammer and drift. Warming the cylinder head will help, as the guides are a tight fit.

12 Oil filters: removal

1 As cleaning the oil filters is part of the routine maintenance the following procedure applies when at least paragraphs 1 to 5 of Section 5 have been completed.
2 Remove the kickstart bolt and pull the kickstart lever clear.
3 Ensure that there is no oil in the engine before removing the eight (C50 and C70 models) or nine screws (C90 model) and the right-hand cover. A deluge of oil will result if this simple precaution is not taken.

11.3a Remove the spring register ...

11.3b ... the inner and outer valve springs ...

11.3c ... and the oil seal and cap

11.5 The valve guide will press out of the head

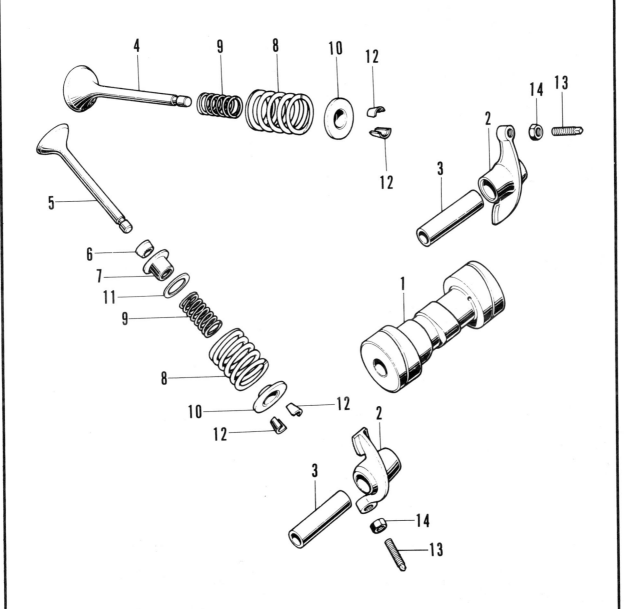

Fig. 1.4. Camshaft and valves - C70 model

1 Camshaft
2 Rocker arm (2 off)
3 Rocker shaft (2 off)
4 Inlet valve
5 Exhaust valve
6 Oil seal
7 Seal cover
8 Outer valve spring (2 off)
9 Inner valve spring (2 off)
10 Spring retainer (2 off)
11 Spring register
12 Collet (4 off)
13 Tappet screw (2 off)
14 Tappet nut (2 off)

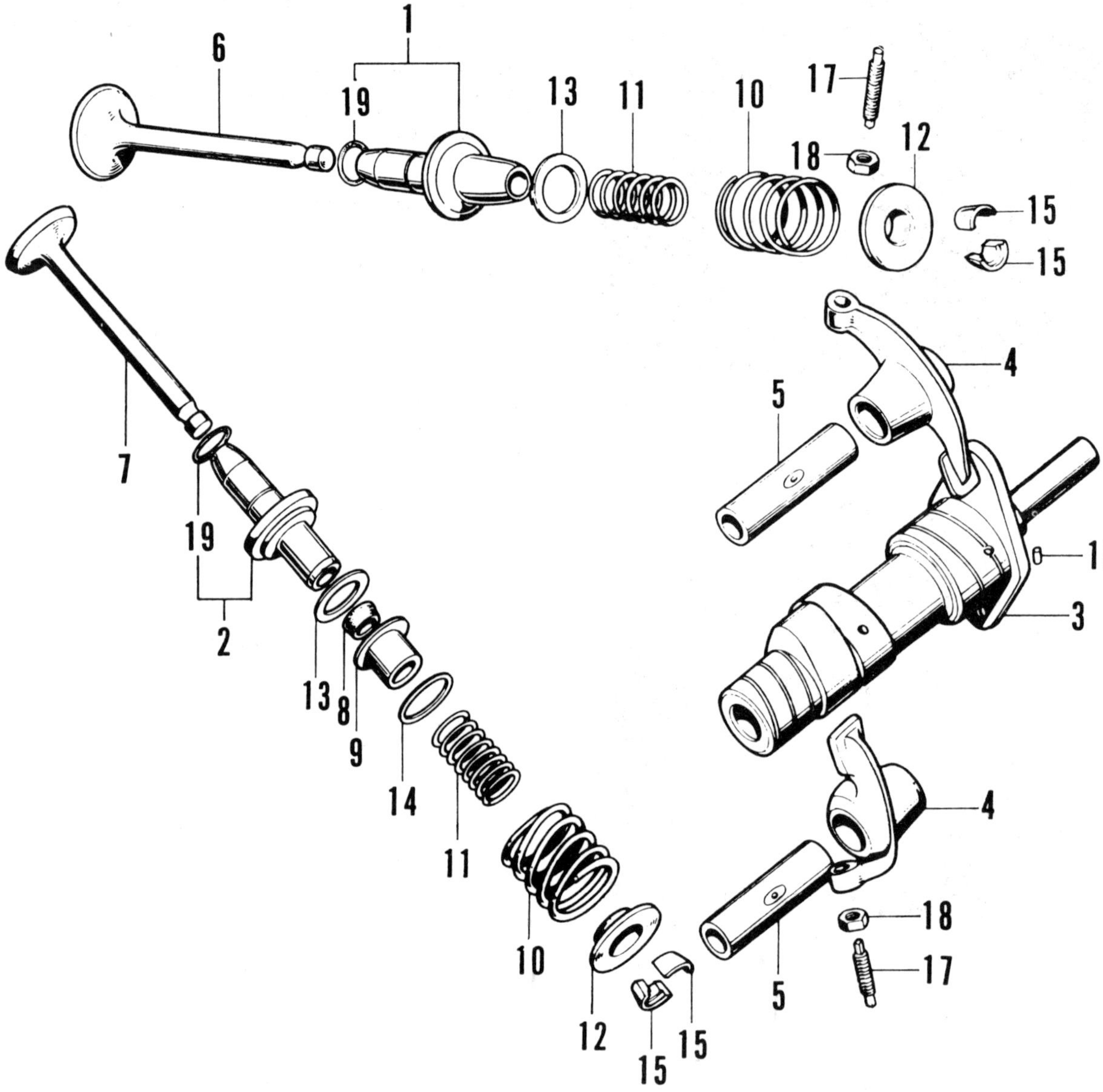

Fig. 1.5. Camshaft and valves - C90 model

1. Inlet valve guide
2. Exhaust valve guide
3. Camshaft
4. Rocker arm (2 off)
5. Rocker shaft (2 off)
6. Inlet valve
7. Exhaust valve
8. Oil seal
9. Seal cover
10. Outer valve spring (2 off)
11. Inner valve spring (2 off)
12. Spring retainer (2 off)
13. Spring register (2 off)
14. Spring register
15. Collet (4 off)
16. Dowel
17. Tappet screw (2 off)
18. Tappet nut (2 off)
19. O-ring (2 off)

Chapter 1: Engine and gearbox

12.4a Remove the clutch operating arm ...

12.4b ... and prise out the camplate

12.5 Remove the clutch outer plate to clean the filter

4 When the cover has been removed, it is probable that an anti-rattle spring will have dropped out of position. This is located between the clutch operating cam plate and the release mechanism, to eliminate chatter. Remove the washer (C50 model only) and the clutch operating lever from its splined shaft, lift off the ball bearing carrier and prise free the camplate from the centre of the clutch.

5 Remove the two (C90 model) or three (C50 and C70 models) screws that retain the clutch outer plate. This is also the centrifugal filter which may have an amount of dirt inside, so care should be taken when removing the plate to ensure that the dirt does not fall into the engine.

6 The filter gauze is located at the bottom of the engine, in a slot probably hidden by the old gasket.

13 Clutch and primary drive: removal

1 As stated in Section 2 of this Chapter, the clutch and primary drive can be removed whilst the engine is still in the frame but Section 12 must be completed first.

2 A special tool is now needed, preferably the Honda service tool, but a suitable equivalent can be made from a piece of tube. The tube is cut to leave two prongs which fit the special sleeve nut, as shown in the accompanying sketch.

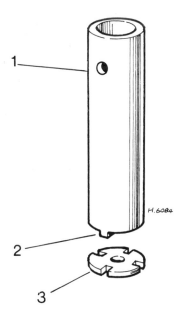

Fig. 1.6. Clutch sleeve tool

1 Holes for tommy bar
2 Two pegs to engage with sleeve nut
3 Sleeve nut

13.3 Remove the sleeve nut

3 Prise out the tab washer, hold the clutch securely and use the special tool to remove the clutch centre nut.
4 The clutch will lift off as a unit and if further work on the clutch is necessary, Chapter 2 will provide all the information required.
5 Slide off from the crankshaft the primary drive pinion, the pinion bearing and the double diameter spacer.
6 Remove the circlip and slide the large primary drive pinion off its shaft.

14 Gearchange mechanism: removal

1 Remove the shouldered bolt and the index arm, with the spring still attached.
2 Remove the bolt in the gearchange drum. Remove the index plate and the four operating pins.
3 The gearchange spindle assembly will now pull clear, provided that the gearchange lever has already been removed.
4 Care should be taken to see that the springs do not fall off and are lost on the floor.

13.4 The clutch lifts off as a complete assembly

13.5a Remove the primary drive pinion ...

13.5b ... the centre bush ...

13.5c ... and the double diameter spacer

13.6a Remove the circlip ...

13.6b ... and slide the gear off its splines

14.1 Remove the pivot bolt and index arm

14.2a Remove the index plate ...

14.2b ... and the four pins

14.3 Slide out the gearchange mechanism

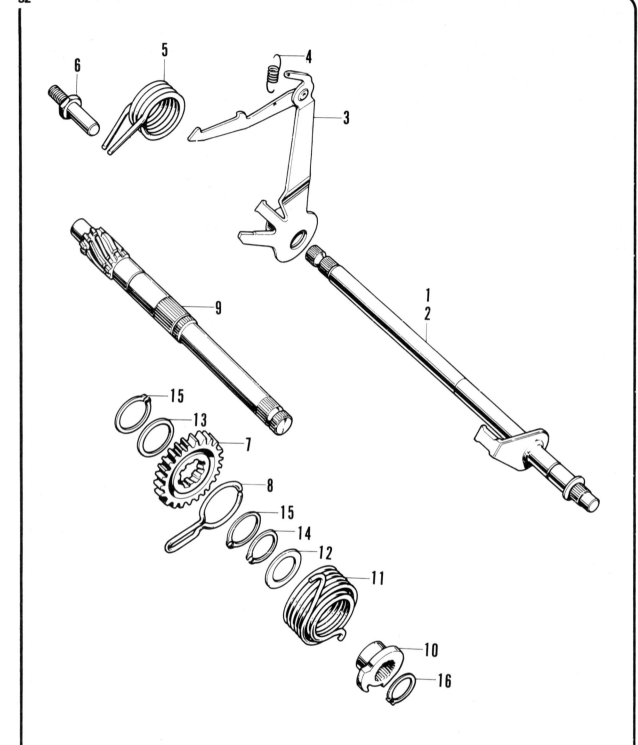

Fig. 1.7. Kickstart and gearchange - C70 model

1 Gearchange spindle assembly
2 Gearchange spindle
3 Gearchange arm
4 Arm spring
5 Return spring
6 Spring stop
7 Pinion
8 Ratchet spring
9 Spindle
10 Spring retainer
11 Return spring
12 Washer
13 Washer
14 Circlip
15 Circlip (2 off)
16 Circlip

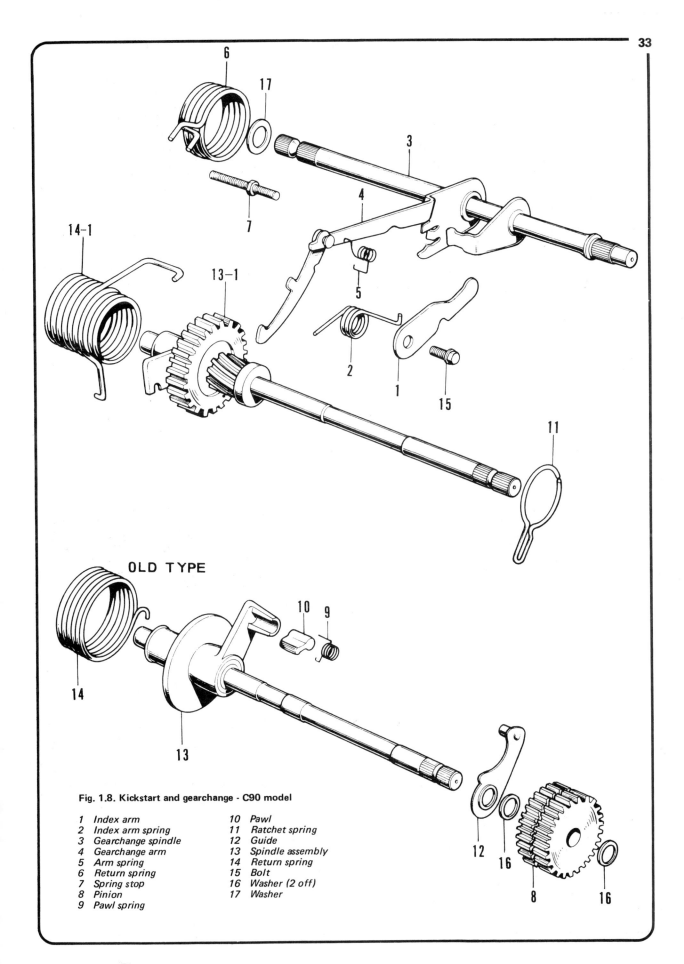

Fig. 1.8. Kickstart and gearchange - C90 model

1 Index arm
2 Index arm spring
3 Gearchange spindle
4 Gearchange arm
5 Arm spring
6 Return spring
7 Spring stop
8 Pinion
9 Pawl spring
10 Pawl
11 Ratchet spring
12 Guide
13 Spindle assembly
14 Return spring
15 Bolt
16 Washer (2 off)
17 Washer

15 Kickstart return spring: removal

C50 and C70 models only

1 Remove the circlip on the kickstart shaft and ease the spring retainer up the shaft whilst releasing the spring from the crankcase. Slide the spring and retainer off the shaft.

16 Final drive sprocket: removal

If the left-hand side cover and the final drive chain have been removed, the sprocket is released by removing the two bolts and the locking plate and pulling the sprocket off its splined shaft.

17 Crankcases: separating

1 If all the necessary components have been removed from the engine as previously described there are only eight screws holding the crankcases together and once these have been removed the right-hand crankcase should lift off with light tapping on the end of the crankshaft and gearshift.

15.1a Release the circlip ...

15.1b ... and slide off the spring and retainer

16.1a Remove the bolts and the locking plate ...

16.1b ... and lift the sprocket clear

17.1 After removing the screws lift the crankcase half clear

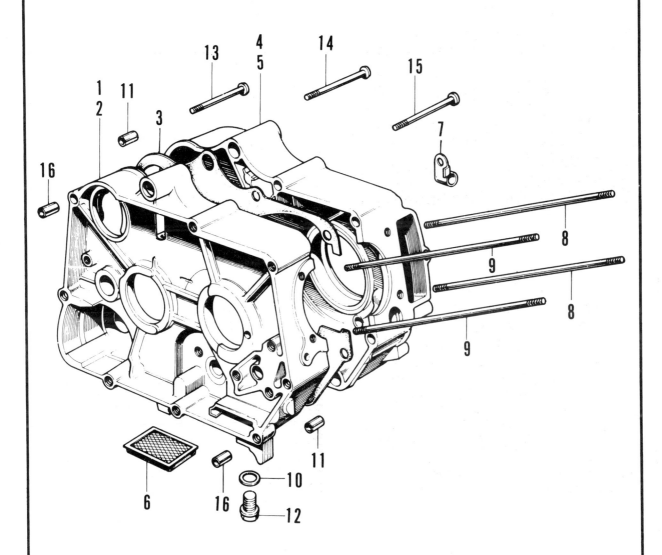

Fig. 1.9. Crankcases - C70 model

1 Right-hand crankcase assembly
2 Right-hand crankcase assembly
3 Crankcase gasket
4 Left-hand crankcase assembly
5 Left-hand crankcase assembly
6 Oil filter
7 Clip
8 Stud (2 off)
9 Stud (2 off)
10 Sealing washer
11 Dowel (2 off)
12 Drain plug
13 Screw
14 Screw (3 off)
15 Screw (4 off)
16 Dowel (2 off)

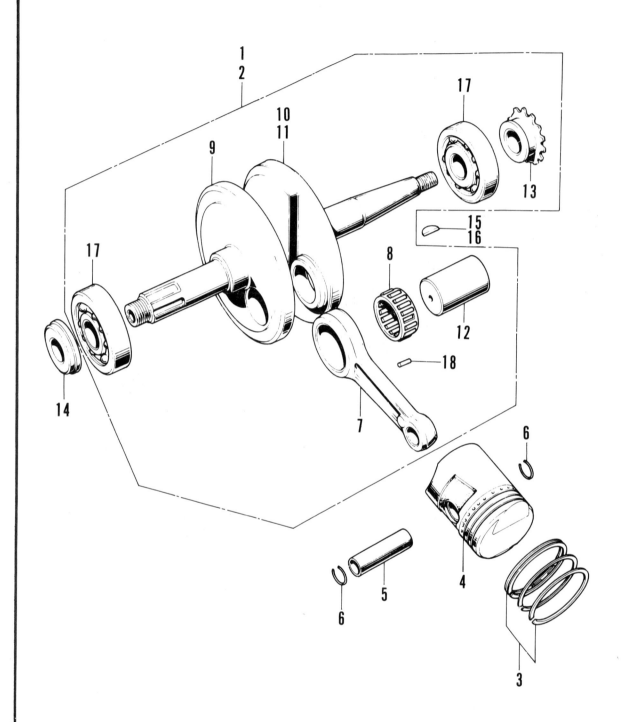

Fig. 1.10. Crankshaft and piston - C70 model

1 Crankshaft assembly
2 Crankshaft assembly
3 Piston ring set
4 Piston
5 Gudgeon pin
6 Circlip (2 off)
7 Connecting rod
8 Roller cage
9 Right-hand crankshaft
10 Left-hand crankshaft
11 Left-hand crankshaft
12 Crank pin
13 Sprocket
14 Spacer
15 Woodruff key
16 Woodruff key
17 Ball bearing (2 off)
18 Big-end roller (22 off)

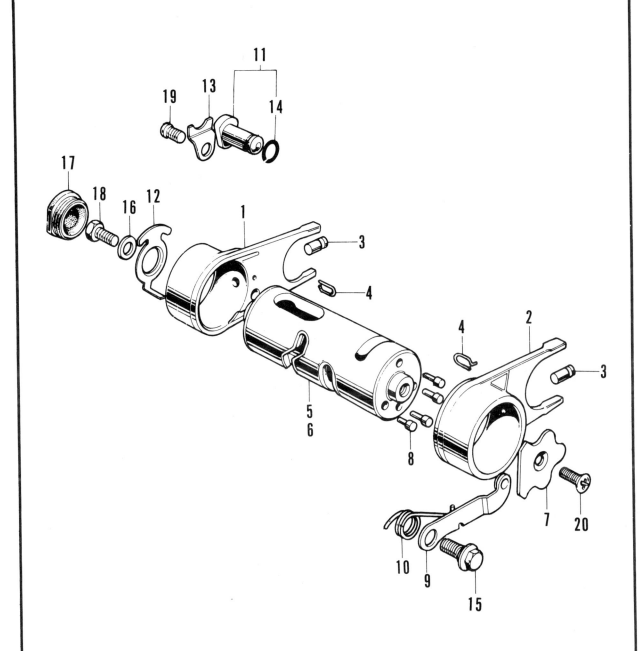

Fig. 1.11. Gearchange drum - C70 model

1 Selector fork
2 Selector fork
3 Selector fork pin (2 off)
4 Pin clip (2 off)
5 Gearchange drum
6 Gearchange drum
7 Index plate
8 Index pin (4 off)
9 Index arm
10 Arm spring
11 Switch assembly
12 Switch contact
13 Retaining plate
14 O-ring
15 Pivot bolt
16 Washer
17 Rubber plug
18 Bolt
19 Screw
20 Screw

Chapter 1: Engine and gearbox

2 Never use the point of a screwdriver to aid the separation of the crankcases. It will cause irreparable damage to the jointing surfaces.
3 There is no need to remove either the oil pump located behind the clutch or the camshaft chain tensioner assembly found at the rear of the flywheel generator. Neither impede the separation of the crankcases.

18 Crankshaft assembly: removal

1 The crankcase bearings are a sliding fit in the steel inserted housings in the crankcase. The crankshaft assembly, complete with bearings, should withdraw from the left-hand case without difficulty, using only light pressure.
2 Note that the camshaft chain will need to be pulled clear of the sprocket on the crankshaft before the crankshaft can be withdrawn.
3 Although it is possible to use an extractor to remove the crankshaft bearings it should be remembered that if the main bearings need replacing the big-end cannot be in the best of condition and a replacement crankshaft assembly is the safest course of action. Note that the cam chain sprocket will need to be removed before the left-hand bearing can be extracted.

19 Kickstart shaft assembly: removal

1 On the C50 and C70 models the kickstart shaft assembly will lift straight out of the crankcase half.
2 On the C90 model, the kickstarter return spring must be unhooked from the crankcase before the kickstart shaft assembly will lift out of the crankcase half. Remove the kickstart return spring.

20 Gear selector drum and gear cluster: removal

1 Remove the rubber blanking plug located at the side of the neutral indicator switch.
2 Remove the 10 mm bolt and washer retaining the gear selector drum in the crankcase (situated adjacent to the neutral indicator). The selector drum can now be withdrawn from the crankcase, together with the selectors and the gear cluster complete.
3 Care should be taken to avoid losing any shims or washers from the ends of the shafts.

21 Neutral indicator switch: removal

1 The neutral indicator switch is retained in position with a metal clamp and a single screw.
2 Remove the screw and clamp and slide the switch out of the crankcase half, taking care not to damage the sealing 'O' ring.

22 Camchain tensioner pulley and oil pump drive: removal

1 To remove the oil pump drive sprocket, hold the sprocket securely and unscrew the oil pump drive shaft. The sprocket can then be pulled clear but a note should be made regarding which way round it is fitted. The C90 model has a one-piece shaft and sprocket.
2 The tensioner ring on the C90 model with the tensioner pulley will lift off once the three bolts and retaining plates have been removed.
3 The tensioner arm on the C50 and C70 models pivots on a single shouldered bolt and removal of this bolt releases the arm.
4 The tensioner push rod if still in the crankcase half can now be pushed out.

18.1 Remove the crankshaft assembly

18.2 Ensure that the camchain has been removed

19.1 Remove the kickstart shaft

20.1 Remove the bolt hidden under the rubber plug

20.2 Remove the gearbox components together

21.1 A screw and plate retain the neutral switch

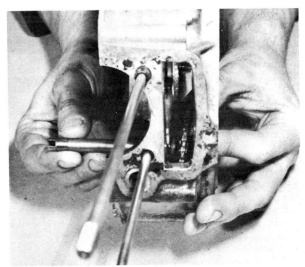
22.1 Unscrew the oil pump driveshaft

22.3 Remove the pivot bolt and tensioner arm

22.4 Ensure the tensioner pushrod is pulled out

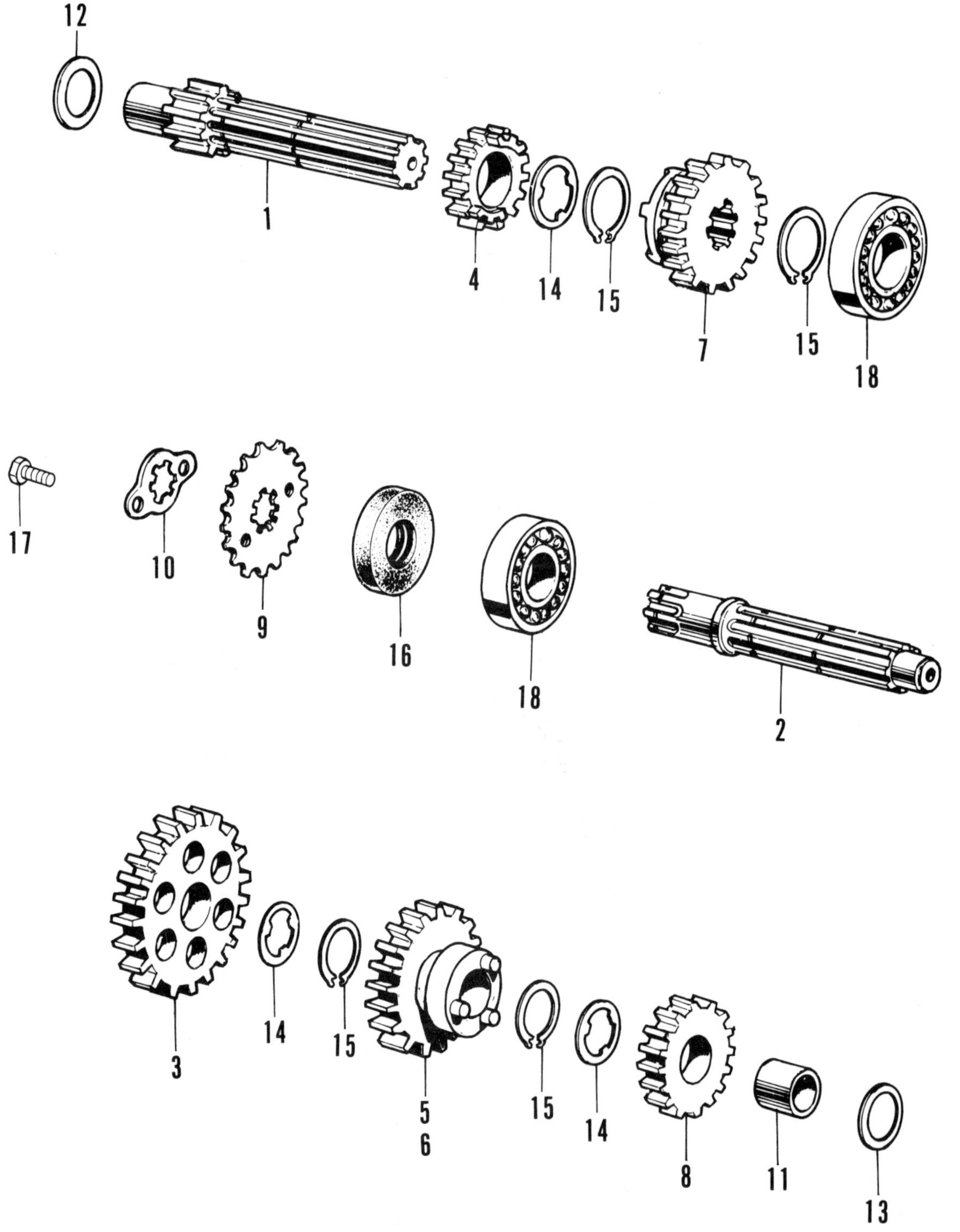

Fig. 1.12. Gearbox components - C70 model

1 Mainshaft	6 Layshaft second gear	11 Bush	16 Oil seal
2 Layshaft	7 Mainshaft top gear	12 Washer	17 Bolt (2 off)
3 Layshaft low gear	8 Layshaft top gear	13 Washer	18 Ball bearing (2 off)
4 Mainshaft second gear	9 Sprocket	14 Splined washer (3 off)	
5 Layshaft second gear	10 Retaining plate	15 Circlip (4 off)	

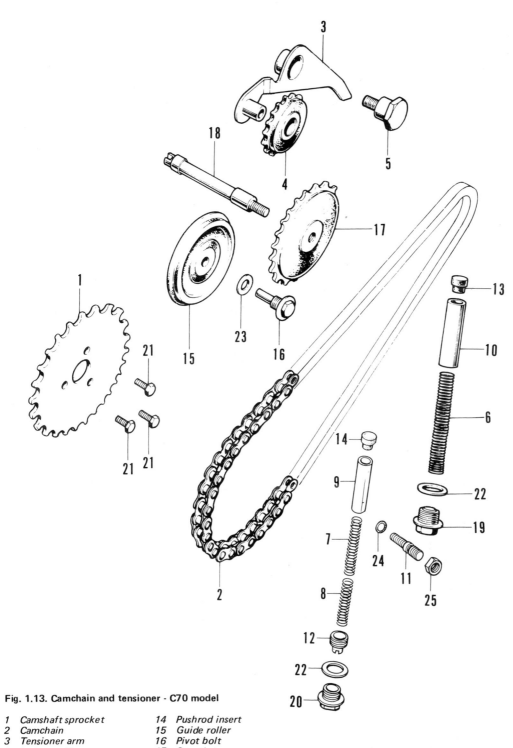

Fig. 1.13. Camchain and tensioner - C70 model

1 Camshaft sprocket
2 Camchain
3 Tensioner arm
4 Jockey pulley
5 Pivot bolt
6 Spring
7 Spring
8 Spring
9 Pushrod
10 Pushrod
11 Adjusting stud
12 Adjusting screw
13 Pushrod insert
14 Pushrod insert
15 Guide roller
16 Pivot bolt
17 Sprocket
18 Oil pump drive
19 Sealing plug
20 Sealing plug
21 Bolt (3 off)
22 Sealing washer
23 Washer
24 O-ring
25 Nut

24.1 Check the condition of the oil seals

25.2 The bearings are a drive fit in the crankcases

27.2 Check the crankshaft assembly for wear

23 Oil pump: removal

1 Remove the three large screws and the oil pump assembly complete will lift clear of the right hand crankcase.
2 The oil pump and the lubrication system are dealt with in Chapter 3 which includes all the necessary information on oil pump operation and renovation.

24 Oil seals: removal

1 Two oil seals are fitted in the left-hand crankcase, located at the gearbox layshaft bearing and gear lever shaft on the C50 and C70 models, whilst the C90 model has the gear lever shaft oil seal in the generator cover. There is also an oil seal on the kickstart shaft bearing in the clutch cover.
2 The oil seals are easily removed by prising them out of position with a screwdriver. Care should be taken to ensure the lip of the bearing housing is not damaged during this operation.

25 Crankshaft and gearbox main bearings: removal

1 The crankshaft bearings will remain on their shafts when the crankshaft assembly is withdrawn from the crankcase. A puller or an extractor will be necessary for their removal as they are a tight fit.
2 The gearbox bearings are a light press fit in the crankcase castings. They can be drifted out of position, using a mandrel of the correct size and a hammer.
3 If necessary, warm the crankcases slightly, to aid the release of the bearings.

26 Examination and renovation: general

1 Before examining the parts of the dismantled engine unit for wear, it is essential that they should be cleaned thoroughly. Use a paraffin/petrol mix to remove all traces of old oil and sludge that may have accumulated within the engine.
2 Examine the crankcase castings for cracks or other signs of damage. If a crack is discovered, it will require professional repair.
3 Examine carefully each part to determine the extent of wear, if necessary checking with the tolerance figures listed in the Specifications section of this Chapter.
4 Use a clean, lint-free rag for cleaning and drying the various components, otherwise there is risk of small particles obstructing the internal oilways.

27 Big-end and main bearings: examination and renovation

1 Failure of the big-end is invariably accompanied by a knock from within the crankcase that progressively becomes worse. Some vibration will also be experienced. There should be no vertical play in the big-end bearing after the old oil has been washed out. If even a small amount of play is evident, the bearing is due for replacement. Do not run the machine with a worn big-end bearing, otherwise there is risk of breaking the connecting rod or crankshaft.
2 It is not possible to separate the flywheel assembly in order to replace the bearing because the parallel sided crankpin is pressed into the flywheels. Big-end repair should be entrusted to a Honda agent, who will have the necessary repair or replacement facilities.
3 Failure of the main bearings is usually evident in the form of an audible rumble from the bottom end of the engine, accompanied by vibration. The vibration will be most noticeable through the footrests.
4 The crankshaft main bearings are of the journal ball type. If wear is evident in the form of play or if the bearings feel rough

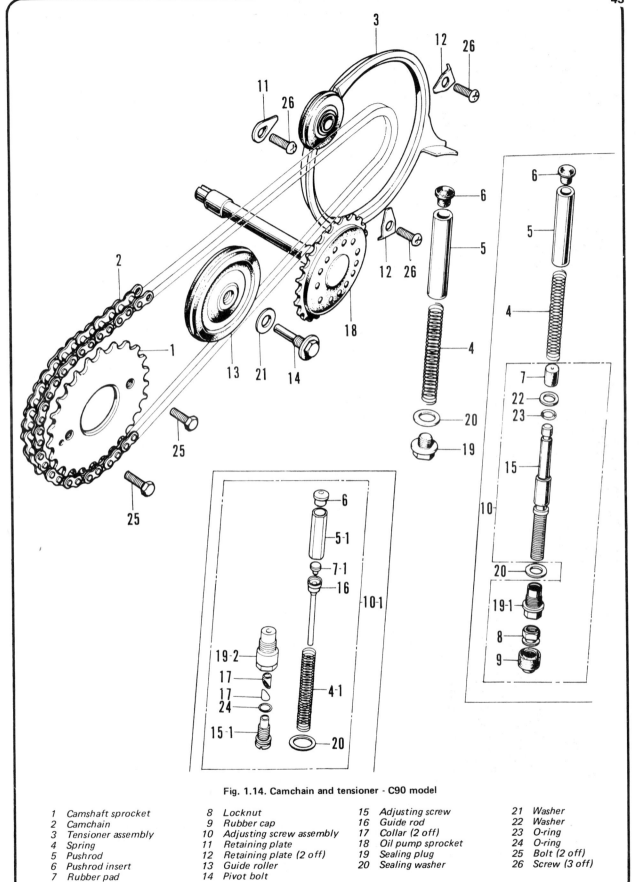

Fig. 1.14. Camchain and tensioner - C90 model

1 Camshaft sprocket	8 Locknut	15 Adjusting screw	21 Washer
2 Camchain	9 Rubber cap	16 Guide rod	22 Washer
3 Tensioner assembly	10 Adjusting screw assembly	17 Collar (2 off)	23 O-ring
4 Spring	11 Retaining plate	18 Oil pump sprocket	24 O-ring
5 Pushrod	12 Retaining plate (2 off)	19 Sealing plug	25 Bolt (2 off)
6 Pushrod insert	13 Guide roller	20 Sealing washer	26 Screw (3 off)
7 Rubber pad	14 Pivot bolt		

as they are rotated, replacement is necessary. To remove the main bearings, if the appropriate service tool is not available, insert two thin steel wedges, one on each side of the bearing, and with these clamped in a vice hit the end of the crankshaft squarely with a rawhide mallet in an attempt to drive the crankshaft through the bearing. When the bearing has moved the initial amount, it should be possible to insert a conventional two or three legged sprocket puller, to complete the drawing-off action.

5   Note that the bottom camshaft chain sprocket must be withdrawn from the left-hand crankshaft before access can be gained to the main bearing. The sprocket is recessed to accommodate a puller.

6   The small end eye should also be checked for wear as the gudgeon pin should be a good fit. The piston should pivot on the gudgeon pin rather than the gudgeon pin rotate in the connecting rod.

28 Cylinder barrel: examination and renovation

1   The usual indications of a badly worn cylinder barrel and piston are excessive oil consumption and piston slap, a metallic rattle that occurs when there is little or no load on the engine. If the top of the bore of the cylinder barrel is examined carefully, it will be found that there is a ridge on the thrust side, the depth of which will vary according to the amount of wear that has taken place. This marks the limit of travel of the uppermost piston ring.

2   Measure the bore diameter just below the ridge, using an internal micrometer. Compare this reading with the diameter at the bottom of the cylinder bore, which has not been subject to wear. If the difference in readings exceeds 0.005 inch it is necessary to have the cylinder rebored and to fit an oversize piston and rings.

3   If an internal micrometer is not available, the amount of cylinder bore wear can be measured by inserting the piston without rings so that it is approximately ¾ inch from the top of the bore. If it is possible to insert 0.004 inch feeler gauge between the piston and the cylinder wall on the thrust side of the piston, remedial action must be taken.

4   Check the surface of the cylinder bore for score marks or any other damage that may have resulted from an earlier engine seizure or displacement of the gudgeon pin. A rebore will be necessary to remove any deep indentations, irrespective of the amount of bore wear, otherwise a compression leak will occur.

5   Check that the external cooling fins are not clogged with oil or road dirt; otherwise the engine will overheat. When clean, a coating of matt cylinder black will help improve the heat radiation.

29 Piston and piston rings: examination and renovation

1   If a rebore is necessary, the existing piston and rings can be disregarded because they will be replaced with their oversize equivalents as a matter of course.

2   Remove all traces of carbon from the piston crown, using a soft scraper to ensure the surface is not marked. Finish off by polishing the crown, with metal polish, so that carbon does not adhere so easily in the future. Never use emery cloth.

3   Piston wear usually occurs at the skirt or lower end of the piston and takes the form of vertical streaks or score marks on the thrust side. There may also be some variation in the thickness of the skirt.

4   The piston ring grooves may also become enlarged in use, allowing the piston rings to have greater side float. If the clearance exceeds 0.004 inch for the two compression rings, or 0.005 inch for the oil control ring, the piston is due for replacement. It is unusual for this amount of wear to occur on its own.

5   Piston ring wear is measured by removing the rings from the piston and inserting them in the cylinder bore using the crown of the piston to locate them approximately 1½ inches from the top of the bore. Make sure they rest square with the bore. Measure the end gap with a feeler gauge; if the gap exceeds 0.010 inch they require replacement, assuming the cylinder barrel is not in need of a rebore.

30 Valves, valve seats and valve guides: examination and renovation

1   After cleaning the valves to remove all traces of carbon, examine the heads for signs of pitting and burning. Examine also the valve seats in the cylinder head. The exhaust valve and its seat will probably require the most attention because these are the hotter running of the two. If the pitting is slight, the marks can be removed by grinding the seats and valves together using fine valve grinding compound.

2   Valve grinding is a simple task, carried out as follows. Smear a trace of fine valve grinding compound (carborundum paste) on the seat face and apply a suction grinding tool to the head of the valve. With a semi-rotary motion, grind in the valve head to its seat, using a backward and forward action. It is advisable to lift the valve occasionally, to distribute the grinding compound evenly. Repeat this operation until an unbroken ring of light grey matt finish is obtained on both valve and seat. This denotes the grinding operation is complete. Before passing to the next operation, make quite sure that all traces of the grinding compound have been removed from both the valve and its seat and that none has entered the valve guide. If this precaution is not observed, rapid wear will take place, due to the abrasive nature of the carborundum base.

3   When deeper pit marks are encountered, it will be necessary to use a valve refacing machine and also a valve seat cutter, set to an angle of 45°. Never resort to excessive grinding because this will only pocket the valve and lead to reduced engine efficiency. If there is any doubt about the condition of a valve, fit a new replacement.

4   Examine the condition of the valve collets and the groove on the valve in which they seat. If there is any sign of damage, new replacements should be fitted. If the collets work loose whilst the engine is running, a valve will drop in and cause extensive damage.

5   Measure the valve stems for wear, making reference to the tolerance values given in the Specifications section of this Chapter. Check also the valve guides, which can be removed by heating the cylinder head in an oven then using a two diameter drift to drive them out of position. The initial diameter of the drift must be a good fit in the valve guide stem. Replace with the new valve guides whilst the cylinder head is still warm.

6   Check the free length of the valve springs against the list of tolerance in the Specifications. If the springs are reduced in length or if there is any doubt about their condition, they should be replaced.

31 Cylinder head: decarbonisation and examination

1   Remove all traces of carbon from the cylinder head and valve ports, using a soft scraper. Extreme care should be taken to ensure the combustion chamber and valve seats are not marked in any way, otherwise hot spots and leakages may occur. Finish by polishing the combustion chamber so that carbon does not adhere so easily in the future. Use metal polish and NOT emery cloth.

2   Check to make sure that valve guides are free from carbon or any other foreign matter that may cause the valves to stick.

3   Make sure the cylinder head fins are not clogged with oil or road dirt, otherwise the engine will overheat. If necessary, use a wire brush.

32 Camshaft, rockers and rocker shafts: examination

1   The cams should have a smooth surface and be free from scuff marks or indentations. It is unlikely that severe wear will be

Chapter 1: Engine and gearbox

encountered during the normal service life of the machine unless the lubrication system has failed or the case hardened surface has broken through.

2 Check the oil groove on the end of the camshaft to ensure it is clean and free from sludge.

3 The internal oilways in the camshaft should also be cleaned and blown through to remove any obstruction.

4 It is unlikely that excessive wear will occur in the rocker arms and rocker shafts, but if it does it will be for the same reasons. A clicking noise from the rocker box is the usual symptom of wear in the rocker components, which should not be confused with the noise that results from excessive tappet clearance. If any shake is present and the rocker arm is loose on its shaft, a new rocker and/or shaft should be fitted.

5 Check the tip of the rocker arm at the point where it bears on the camshaft. If signs of cracking, scuffing or breakthrough are found in the case hardened surface, fit a new replacement. Check also the condition of the thread on the tappet, the rocker arm and the locknut.

33 Camshaft chain tensioner and sprockets: examination

1 An oil damped camshaft chain tensioner is employed, to fulfill the dual function of controlling the chain tension at high engine speeds and eliminating mechanical noise. A compression spring and pushrod within a guide provides the tension by bearing on one end of a pivoting arm which carries a jockey pulley on the other. The jockey pulley engages with the top run of the chain. The guide containing the spring and pushrod floods with oil when the engine is running, to provide the necessary damping medium.

2 The chain tensioner spring should have a free length of 3.04 inches. It should be replaced if the free length is reduced below 2.89 inches. To gain access to the spring and pushrod assembly, remove the 14 mm bolt that screws at an angle into the base of the left-hand crankcase.

3 Adjustment of the tensioner is automatic. Check that the tensioner is operating correctly when the left-hand flywheel generator cover and flywheel generator are removed.

4 The camshaft is of the endless variety and should not contain a split link.

5 It is unlikely that the camchain sprockets will need renewing unless the teeth have been damaged or broken.

6 Clean the sprockets so that any timing marks are easily identified.

34 Gearchange mechanism: examination

1 Examine the mechanism for any signs of damage, renewing any of the springs which may have become weak or broken.

2 Check for wear on the gearchange lever pawls as this can cause missed gearchanges.

35 Kickstart assembly: examination

1 Give the kickstart assembly a close visual inspection for signs of wear or damage such as broken or chipped teeth, removing the necessary circlips if dismantling for renewal of any parts.

2 Examine the kickstart return spring for weakness or damage. This component is often overlooked, even though it is tensioned every time the kickstart is depressed. It is best to renew as a precaution during a major overhaul, to prevent a further stripdown later.

36 Primary drive gears: examination

Both primary drive gears should be examined closely to ensure that there is no damage to the teeth. The depth of mesh is predetermined by the bearing locations and cannot be adjusted.

37 Gear selector drum and gear cluster: examination

1 This group of components was removed from the crankshaft as a unit, with care being taken to avoid losing any shims or washers from the ends of the shafts. The parts fall naturally into three sub-assemblies; the selector drum, the mainshaft and the layshaft.

2 The selector drum sub-assembly should be examined to ensure that the selector forks will slide easily on the drum without too much play. Check that the selector forks are not bent or excessively worn. To renew either selector fork, remove the spring clip and the cam track follower and slide the selector fork clear. When reassembling, ensure that the selector fork is fitted the right way round.

3 The mainshaft and layshaft sub-assemblies should be examined closely for any signs of wear or damage such as

33.5 Check the sprocket for worn or broken teeth

35.1 Check the condition of the ratchet spring

Chapter 1: Engine and gearbox

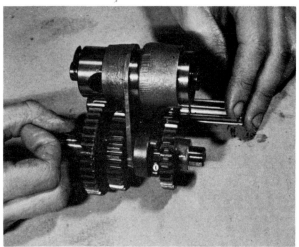

37.1 Note the relative positions before dismantling the gearbox components

37.2 Check the selector forks for wear or damage

37.3a Check for wear on the mainshaft components.

37.3b ... and the layshaft components ...

37.3c ... and ensure the circlips are properly seated on reassembly

broken or chipped teeth, worn dogs and damaged or worn splines. Renew any parts found unserviceable as they cannot be reclaimed. To renew any parts, removal of the circlips, washers and gears until the defective part is reached is straightforward and if the parts are laid out in sequence, reassembly should present no problems.

38 Engine reassembly: general

1 Before reassembly is commenced, the various engine and gearbox components should be thoroughly clean and placed close to the working area.
2 Make sure all traces of the old gaskets have been removed and the mating surfaces are clean and undamaged. One of the best ways to remove old gasket cement is to apply a rag soaked in methylated spirit. This acts as a solvent and will ensure the cement is removed without resort to scraping and the consequent risk of damage.
3 Gather together all the necessary tools and have available an oil can filled with clean engine oil. Make sure all the new gaskets and oil seals are to hand; nothing is more frustrating than having to stop in the middle of a reassembly sequence because a vital gasket or replacement has been overlooked.

Chapter 1: Engine and gearbox

4 Make sure the reassembly area is clean and that there is adequate working space. Refer to the torque and clearance settings wherever they are given. Many of the smaller bolts are easily sheared if they are over-tightened. Always use the correct size screwdriver bit for the crosshead screws and never an ordinary screwdriver or punch.

39 Engine reassembly: fitting the bearings and oil seals to the crankcases and clutch cover

1 Before fitting any of the crankcase bearings make sure that the bearing housings are scrupulously clean and that there are no burrs or lips on the entry to the housings. Press or drive the bearings into the cases using a mandrel and hammer, after first making sure that they are lined up squarely. Warming the crankcases will help when a bearing is a particularly tight fit.
2 When the bearings have been driven home, lightly oil them and make sure they revolve smoothly. This is particularly important in the case of the main bearings.
3 Using a soft mandrel, drive the oil seals into their respective housings. Do not use more force than is necessary because the seals damage very easily.
4 Lightly oil all the other moving parts as a prelude to reassembly. This will ensure all working parts are lubricated adequately during the initial start-up of the rebuilt engine.

40 Engine reassembly: refitting the oil pump

1 Reference to Chapter 3 will fully explain the operation and renovation of the oil pump so that it is ready to fit to the crankcase as a sub-assembly.
2 Smear a very thin film of jointing compound such as Golden Hermatite onto the crankcase face and stick the gasket in position. Do not use an excessive amount of jointing compound as serious engine damage can result if any of the oilways are blocked.
3 Fit the oil pump in position and secure it with three screws. Check that it is free to rotate.

41 Engine reassembly: refitting the camchain tensioner pulley and oil pump drive

1 On the C50 and C70 models, fit the tensioner arm and its shouldered pivot bolt.
2 On the C90 model, fit the tensioner ring and secure it in position with the three retaining plates and bolts.
3 Feed the oil pump drive sprocket into the crankcase ensuring that it fits the right way round, and screw in the drive shaft. Hold the sprocket securely and fully tighten the shaft.

42 Engine reassembly: refitting the neutral indicator switch

1 Ensure that the sealing 'O' ring on the neutral indicator switch is in good condition before pushing the switch into the crankcase.
2 Fit the switch retaining plate and screw.

43 Engine reassembly: replacing the gear selector drum and gear cluster

1 Place the left-hand crankcase on wooden blocks or an engine stand so that the inner side faces upwards.
2 Engage the selector forks in their respective positions, with the sliding dog on the layshaft and second gear on the mainshaft. Viewed endwise the lower of the two selector forks engages with the sliding dog on the layshaft and the upper fork with the sliding second gear pinion on the mainshaft.

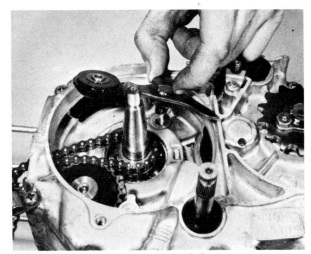

41.1 Refit the tensioner arm and pivot bolt

41.3 Ensure that the sprocket is the right way round

42.1 Check that the O-ring is in good condition

3 Holding the complete gear assembly in the right-hand, locate the layshaft in the journal ball bearing, the mainshaft in the plain bush and the tapered end of the selector drum in its housing. The 10 mm bolt and washer that hold the selector drum can then be fitted into the underside of the crankcase half. Tighten this bolt and later fit the rubber blanking plug.

44 Engine reassembly: replacing the kickstart shaft assembly

1 The kickstart shaft will feed into the crankcase half but on the C90 model the kickstart return spring will need retensioning before the shaft will seat home properly.
2 Ensure that the kickstart friction spring is properly located in the crankcase half.

45 Engine reassembly: replacing the crankshaft assembly

1 Fit the crankshaft assembly in the left-hand crankcase with the splined mainshaft uppermost. Make sure the connecting rod clears the aperture for the cylinder barrel spigot. It may be

43.3a Reassemble the gearbox cluster ...

43.3b ... and retain the selector drum with this bolt

44.2 Refit the kickstart shaft with ratchet spring in cast groove

45.1 Refit the crankshaft assembly

46.3 Refit the crankcase half and ensure all the shafts still rotate

necessary to tap the assembly into position, if the crankshaft journal ball bearing is a tight fit in the steel outer ring.

46 Engine reassembly: rejoining the crankcases

1 Smear the joint face with Golden Hermatite or other jointing compound and stick the gasket to the crankcase joint.
2 Ensure that the crankcase dowels are in position.
3 Lower the other crankcase half into position. Gentle tapping may be required to fit the two halves together as the bearings and dowels are a tight fit.
4 Rotate all the shafts to ensure that they will turn and that no binding occurs, especially the oil pump drive shaft, to ensure that the slot engages with the oil pump.
5 Excessive force should not be used as this shows something has been wrongly assembled or is out of alignment.
6 Replace the eight screws which hold the crankcase together.

47 Engine reassembly: refitting the final drive sprocket

1 Now is an ideal time to fit the rubber blanking plug over the end of the gear selector drum.
2 The sprocket is pushed on the splines followed by the locking plate. The locking plate is turned in the groove and the two bolts tightened to clamp the plate to the sprocket.

48 Engine reassembly: replacing and tensioning the kickstart return spring (C50 and C70 models only)

1 Rotate the kickstart shaft clockwise until it will go no further.
2 Fit the spring and retainer onto the shaft and engage the splines by gently easing the shaft anti-clockwise to ensure that the spring retainer hits the return stop.
3 Wind the kickstart return spring round until it hooks into the crankcase. Refit the retaining circlip.

49 Engine reassembly: refitting the gearchange mechanism

1 Check the condition of the pawl spring and the return spring on the gearchange shaft assembly before assembling it into the crankcase.

47.1 Refit the rubber plug

47.2a Refit the sprocket ...

47.2b ... relocate the retaining plate ...

47.2c ... and tighten the two bolts

Chapter 1: Engine and gearbox

48.3 Replace the circlip to hold the spring retainer

49.1 Ensure the return spring fits round the spring stop

49.3a Refit the four pins ...

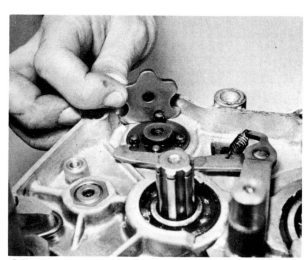

49.3b ... and retain them with the index plate

2 Grease the oil seal and carefully feed the gearchange shaft into position, ensuring that the splines do not damage the oil seal. An alternative method is to fit a small plastic bag over the shaft for feeding through the oil seal.
3 Refit the four small pins and the index plate onto the end of the gearchange drum, noting that the index plate is seated properly in the correct position.
4 Check the condition of the index arm spring and then refit the index arm with the shouldered bolt.

50 Engine reassembly: reassembling the clutch and primary drive

1 The large primary drive pinion is fitted on the splines of the mainshaft and retained with a circlip.
2 Fit the double diameter spacer, the pinion bearing and the small drive pinion onto the crankshaft.
3 Reference to Chapter 2 will fully explain the operation and renovation of the clutch, so that it is ready to fit the crankshaft as a sub-assembly.
4 Fit the clutch sub-assembly, the tab washer and the special nut to the crankshaft.
5 Hold the clutch securely and tighten the sleeve nut to a torque of 27.5 to 32.5 ft lb as recommended in the Specifications section of this Chapter. Prise the tab washer into one of the slots in the sleeve nut.

51 Engine reassembly: replacing the oil filters

1 Smear a thin film of jointing compound such as Golden Hermatite on the outer face of the clutch and stick the gasket into position.
2 Refit the clutch outer plate to form the centrifugal filter and retain it with two screws (on the C90 model) or three screws (C50 and C70 models).
3 Check that the orifice in the clutch camplate is clean and the pressure relief spring is in good condition.
4 Fit the orifice and spring into the camplate and fit the camplate into the clutch outer plate.
5 Position the anti-rattle spring and locate the ball bearing carrier onto the spring.
6 Refit the clutch operating arm onto its splines, ensuring that the arm points towards the centre of the clutch. The C50 model has an additional washer fitted on the shaft.

49.4 Refit the index arm and pivot bolt

50.1 Retain the gear with the circlip

50.2a Refit the double diameter spacer and the bush ...

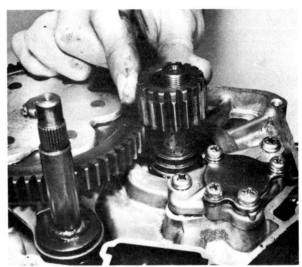
50.2b ... and slide the gear pinion into position

50.4 Ensure the clutch is properly located on its splines and the primary drive pinion

50.5 Prise the tab washer into one of the slots when the nut is fully tightened

51.4 Refit the camplate and oil orifice

51.5 Fit the spring and the ball retainer

51.6 Refit the operating arm

51.7 Slide the filter gauze into its slot

52.2a Reassemble the valve ...

52.2b ... the oil seal ...

Chapter 1: Engine and gearbox

7 Slide the clean filter gauze each into its slot in the bottom of the crankcase.

8 Smear the joint face with Golden Hermatite or other jointing compound and stick the gasket to the crankcase. Ensure that the two dowels are fitted correctly.

9 Slacken the clutch adjustment stud and locknut and lower the clutch cover into position, ensuring that the ball bearing carrier is not disturbed.

10 Gentle tapping may be required to ensure that the cover is fully home on the dowels. Care should be taken to ensure that the oil seal on the kickstart shaft is not damaged.

11 Replace the eight screws (on the C50 and C70 models) or nine screws (C90 model) which hold the cover but do not fully tighten the screw that carries the spark plug lead clip.

12 Readjust the clutch release mechanism and tighten the locknut as described in Chapter 2.6.

13 The kickstart lever and bolt can be replaced at this stage, but it is customary to refit these components once the engine is in the frame.

52 Engine reassembly: replacing the valves and valve guides

1 When reassembling the valve guides into the cylinder head use a drift that fits the guide, such as the Honda service tool or use a long bolt and spacers and draw the guide into the head. It is possible to fractionally close the bore of a valve guide with hammering so that a valve will not slide freely - even though new parts have been used. Do not forget the 'O' ring seal.

2 The valve is fitted into the guide. The oil seal, oil seal cover and bottom spring register are fitted to the exhaust valve only, then the inner and outer valve springs and the spring register are clamped with a small valve spring compressor and the two half collets fitted. When removing the compressor ensure that the half collets are seating correctly.

3 The above procedure applies to both valves. The exhaust valve guide oil seal should be checked for damage if the engine has an oily exhaust.

53 Engine reassembly: replacing the camshaft and rocker arms

1 The camshaft of the C50 and C70 models slides into the cylinder head once the cam lobes are lined up with the cutouts in the cylinder head. The C90 model camshaft is fitted after the cylinder head is attached to the engine.

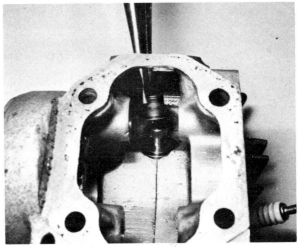

52.2c ... the oil seal cap ...

52.2d ... the inner and outer valve springs ...

52.2e ... the spring register ...

52.2f ... and retain it all with two collets

53.1 Line up the cam lobes with the cutaways in the head

2 Slacken the tappet adjusting screws, position each rocker arm and slide the rocker pin into place, ensuring that the extraction thread is outermost.
3 On the C50 and C70 models the tappet adjustment can be made now but the operation is described later in this Chapter, when the engine is in the frame.

54 Engine reassembly: refitting the piston and cylinder barrel

1 Raise the connecting rod to its highest point and pad the mouth of the crankcase with clean rag as a precaution against displaced parts falling into the crankcase.
2 Assemble the piston on the connecting rod, with the arrow on the piston crown facing downwards.
3 Lightly oil and fit the piston onto the connecting rod by inserting the gudgeon pin. Replace the circlips that retain the gudgeon pin making doubly sure that they are correctly seated in their grooves. Always renew the circlips as it is false economy to re-use the originals.
4 Thread the cam chain into position and ensure it seats properly onto the sprockets. Check that the two dowels are

53.2a Reposition the rocker arms ...

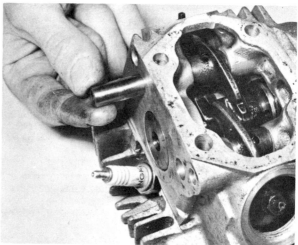

53.2b ... and slide the rocker pins into position

54.2 Refit the piston ensuring circlips are seated correctly

54.4 Thread the camshaft chain into position

properly located on the holding down studs.

5 The cylinder base gasket should be stuck in position with Golden Hermatite or other jointing compound, and the oil seal for the oil hole pushed into position.

6 The cylinder barrel should be fed onto the holding down studs. A piece of wire can be used to hook the cam chain through the cam chain tunnel.

7 The barrel should be lightly oiled and pushed further down the studs until the piston starts to enter the bore. The piston rings can then be compressed and fed into the bore.

8 When all three piston rings are in the bore the padding in the crankcase can be removed and the barrel slid down the studs to locate with the two dowels.

9 On the C50 and C70 models the cylinder barrel is retained in position by fitting the bolt on the left-hand side of the engine, finger tight.

10 Refit the guide roller into the camchain tunnel and retain it with the bearing bolt.

55 Engine reassembly: replacing the cylinder head and timing the valves (C50 and C70 models only)

1 Before the cylinder head can be fitted to the engine it must be fully assembled. Although it may appear possible to replace the rockers when the cylinder head is bolted down, this is not so in practice. The rocker spindles are retained by the long holding down studs that pass through the cylinder head and cannot be removed or replaced unless the cylinder head is lifted.

2 Refit the dowels on the holding down studs and fit a new cylinder head gasket and its associated 'O' ring for sealing the oil passageways.

3 To ease later assembly, ensure that the piston is at top dead centre (TDC) and the camshaft is also positioned such that the cam lobes point downward, the corresponding position to top dead centre on the compression stroke. Fit the camshaft sprocket into the camchain so that the 'O' mark is positioned at the top.

4 Lower the cylinder head onto the studs and feed the camshaft sprocket into the camchain tunnel. Locate the cylinder head on the two dowels and push the sprocket onto the spigot on the end of the camshaft.

5 Fit the holding down bolt on the left-hand side of the engine, finger tight.

6 Check that the engine is still at top dead centre and with the lower run of the camchain taut, that the 'O' mark on the

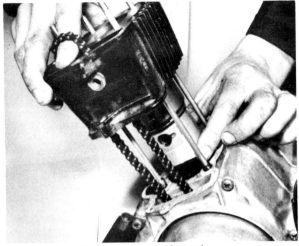
54.6 Slide the cylinder barrel down the studs

54.10 Refit the guide roller in the barrel

55.3 Fit a new gasket and seals and the sprocket

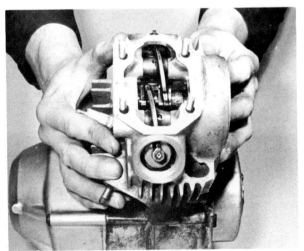
55.4a Slide the cylinder head down the studs ...

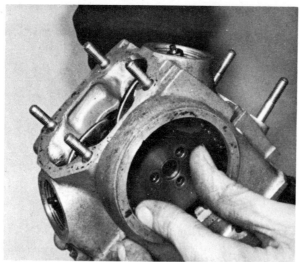
55.4b ... and fit the sprocket onto the camshaft

55.6a Line up the 'T' mark with the crankcase mark ...

55.6b ... and the 'O' mark with the cylinder head mark for correct valve timing

55.7a Reassemble the tensioner pushrod ...

55.7b ... and spring and refit the sealing plug

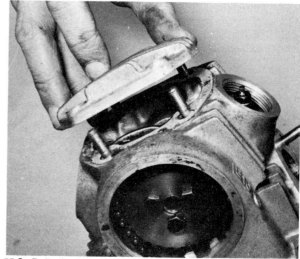
55.8a Refit the top cover ...

Chapter 1: Engine and gearbox

sprocket lines up with the notch on the cylinder head. The camshaft should be in the correct position for the three bolts to be fitted, retaining the sprocket.

7 Slide the camchain tensioner plunger and pressure spring into the crankcase and fit the sealing plug, after ensuring that the sealing washer is in good condition. Recheck the timing to ensure that it is correct.

8 The top cover gasket should be stuck in position with Golden Hermatite or other jointing compound and the top finned cover fitted, so that the arrow between the central fins points towards the exhaust valve.

9 Fit the nuts and washers ensuring that the domed nuts and sealing washers are fitted in their correct positions, and pull down evenly until the recommended torque settings are achieved (6.5 - 8.7 ft lb). Always tighten in a diagonal sequence, an essential requirement because an alloy cylinder head will distort easily. There is a separate bolt on the left-hand side of the cylinder head, just below the circular camshaft sprocket aperture, and below this, a bolt at the base of the cylinder barrel. Both must be tightened, as they are only finger tight at present.

10 Stick the side cover gaskets into position, again using Golden Hermatite, and bolt the finned cover then the circular cover into place. Refit the spark plug.

C90 model only

11 Before the cylinder head can be fitted to the engine, the rocker arms must be assembled. Although it may appear possible to replace the rockers when the cylinder head is bolted down, this is not so in practice. The rocker spindles are retained by the long holding down studs that pass through the cylinder head and cannot be removed or replaced unless the cylinder head is lifted.

12 Refit the dowels on the holding down studs and fit a new cylinder head gasket, oil feed seal and camchain tunnel seal.

13 To ease later assembling, ensure that the piston is at top dead centre (TDC) and fit the camshaft sprocket into the camchain so that the 'O' mark is positioned at the top.

14 Lower the cylinder head onto the studs and feed the camshaft sprocket into the camchain tunnel. Locate the cylinder head on the two dowels.

15 Feed the camshaft into the cylinder head through the sprocket and line the dowel pin up with 'O' mark, which should be in line with the notch on the cylinder head. Fit and finger tighten the two bolts securing the sprocket to the camshaft.

16 Slide the camchain tensioner plunger and pressure spring into the crankcase and fit the sealing plug, after ensuring that the sealing washer is in good condition. Recheck the timing to ensure that it is correct.

17 The top cover gasket should be stuck in position with Golden Hermatite or other jointing compound and the top finned cover fitted so that the arrow between the central fins points towards the exhaust valve.

18 Fit the nuts and washers, ensuring that the domed nuts and sealing washers are fitted in their correct positions, and pull down evenly until the recommended torque settings are achieved (6.5 - 8.7 ft lb). Always tighten in a diagonal sequence, an essential requirement because an alloy cylinder head will distort easily.

19 Fully tighten the camshaft bolts.

20 Stick the side gaskets into position, again using Golden Hermatite, and bolt first the finned side cover then fix the contact breaker base into place with three screws. Ensure that the oil seal in the base is in good condition and undamaged.

21 Refit the dowel pin and the automatic advance mechanism and replace the bolt in the centre of the camshaft. Check the condition of the springs and the smooth action of the advance mechanism, renovating if necessary.

22 Refit the contact breaker assembly complete with the back plate and retain it with two screws, lining up with the scribe marks made inside during the dismantling stage. Reconnect the lead wire. If the contact breaker needs adjusting refer to Chapter 4 for full details. Always check the ignition timing, even if the scribe marks made earlier are in alignment with one another.

23 Stick the cover gasket into position again using Golden Hermatite and refit the cover and two screws. Refit the spark plug.

56 Engine reassembly: adjusting the tappets

1 The tappets should be adjusted to 0.002 inch clearance when the engine is cold and the piston is at top dead centre (TDC) on the compression stroke.

2 To adjust the tappets, slacken the locknut at the end of the rocker arm and turn the square-ended adjuster until the clearance is correct, as measured by a feeler gauge. Hold the square-ended adjuster firmly when the locknut is tightened, otherwise it will move and the adjustment will be lost.

3 After completing the adjustment to both valves, refit and tighten the rocker box caps, using new 'O' ring seals. Use a spanner that is a good fit otherwise the caps will damage easily.

55.8b ... with the arrow pointing towards the exhaust valve

55.10 Refit the side covers

Chapter 1: Engine and gearbox

56.3 Refit the tappet covers

57 Engine reassembly: replacing the flywheel generator

C50 and C70 models only

1 Fit the two small 'O' rings that seal the stator plate screws in their counterbores, and fit the Woodruff key into the crankshaft.
2 Ensure that the central oil seal and the large 'O' ring on the outside diameter of the stator plate are in good condition and undamaged before fitting the plate in position. Secure it with the two screws, aligning the scribe marks made when the stator plate was removed, and fit the rubber grommets on the wires into their respective cutouts.
3 Reconnect the green/red striped wire to the neutral indicator switch.
4 Before fitting the flywheel rotor, place a few drops of light oil on the felt wick which lubricates the contact breaker cam in the centre of the flywheel rotor.
5 It is advisable to check also whether the contact breaker points require attention at this stage, otherwise it will be necessary to withdraw the flywheel rotor again in order to gain access. Reference to Chapter 4, will show how the contact breaker points are renovated and adjusted.

57.1 Check that all the seals are in good condition

57.2 Refit the stator plate

57.8 Refit the rotor

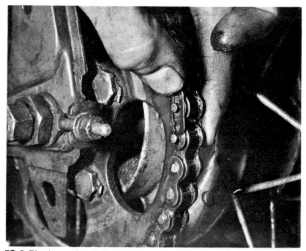

58.2 Fit the spring link with the closed end facing the direction of chain travel

Chapter 1: Engine and gearbox

6 Check the condition of the springs and the smooth action of the automatic advance mechanism, renovating if necessary.
7 Feed the rotor onto the crankshaft so that the slot lines up with the Woodruff key. The rotor may have to be turned to clear the heel of the contact breaker before it will slide fully home.
8 The washer and rotor nut can now be fitted and the nut fully tightened, to the specified torque of 23.9 - 27.5 ft lb.
9 Refit the flywheel cover and secure it with three screws only if the engine is in the frame. Check the ignition timing, to verify it is correct.

C90 model only

1 Fit the rotor and secure it with its bolt and washer, tightening the bolt to the specified torque of 23.9 - 27.5 ft lb.
2 Fit the stator coils and secure them with the two screws, and fit the rubber grommets on the wires into the cutouts.
3 Reconnect the green/red striped wire to the neutral indicator switch.
4 Ensure that the two dowels are fitted correctly and smear Golden Hermatite or other jointing compound onto the crankcase joint face. Stick the gasket in position and fit the generator cover securing it with eight screws. Check the gearchange lever oil seal for any sign of deterioration or damage before fitting the cover.
5 Refit the sprocket cover and two screws if the engine is in the frame.
6 If the inspection cover has been removed, stick the gasket in position, again using Golden Hermatite, refit the cover and retain with the three screws. The condition of the 'O' rings on the three screws should be checked for any sign of deterioration or damage.

58 Refitting the engine/gearbox unit in the frame

1 Follow in reverse the procedure given in Section 5 of this Chapter with the following points borne in mind:
2 Check that the final drive chain link is fitted the correct way round. The closed end of the spring link should lead as the chain rotates.
3 Ensure that a new copper/asbestos joint ring is used in the exhaust port as a leakproof joint is essential for the correct running of the engine.
4 Ensure that the sealing washer of the drain plug is in good condition and fully tighten the drain plug. Refill the engine unit with oil of the recommended viscosity, to the correct level.

59 Starting and running the rebuilt engine

When the initial start-up is made, run the engine gently for the first few minutes in order to allow the oil to circulate throughout all parts of the engine. Remember that if a number of new parts have been fitted or if the engine has been rebored, it will be necessary to follow the original running-in instructions so that the new parts have ample opportunity to bed-down in a satisfactory manner. Check for oil leaks and/or blowing gaskets before the machine is run on the road.

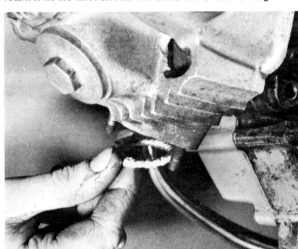

58.3 Always use a new gasket for a leaktight joint

58.4 Refill the engine with the correct oil

60 Fault diagnosis: engine

Symptom	Cause	Remedy
Engine does not start	Lack of compression	
	Valve stuck open	Adjust tappet clearance.
	Worn valve guides	Renew.
	Valve timing incorrect	Check and adjust.
	Worn piston rings	Renew.
	Worn cylinder	Rebore.
	No spark at plug	
	Fouled or wet spark plug	Clean.
	Fouled contact breaker points	Clean.
	Incorrect ignition timing	Check and adjust.
	Open or short circuit in ignition	Check wiring and cut-out switch.
	No fuel flowing to carburettor	
	Blocked fuel tank cap vent hole	Clean.
	Blocked fuel tap	Clean.

Symptom	Cause	Remedy
	Faulty carburettor float valve	Renew.
	Blocked fuel pipe	Clean.
Engine stalls whilst running	Fouled spark plug or contact breaker points	Clean.
	Ignition timing incorrect	Adjust.
	Blocked fuel line or carburettor jets	Clean.
Noisy engine	Tappet noise:	
	Excessive tappet clearance	Check and reset.
	Weakened or broken valve spring	Renew springs.
	Knocking noise from cylinder:	
	Worn piston and cylinder noise	Rebore cylinder and fit oversize piston.
	Carbon in combustion chamber	Decoke engine.
	Worn gudgeon pin or connecting rod small end	Renew.
	Cam chain noise	Adjust.
	Stretched cam chain (rattle)	Renew chain.
	Worn cam sprocket or timing sprocket	Renew sprockets.
Engine noise	Excessive run-out of crankshaft	Renew.
	Worn crankshaft bearings (rumble)	Renew.
	Worn connecting rod or big end (knock)	Renew flywheel assembly.
	Worn transmission splines	Renew.
	Worn or binding transmission gear teeth	Renew gear pinions.
Smoking exhaust	Too much engine oil	Check oil level and adjust as necessary.
	Worn cylinder and piston rings	Rebore and fit oversize piston and rings.
	Worn valve guides	Renew.
	Damaged cylinder	Renew cylinder barrel and piston.
Insufficient power	Valve stuck open or incorrect tappets adjustment	Re-adjust.
	Weak valve springs	Renew.
	Valve timing incorrect	Check and reset.
	Worn cylinder and piston rings	Rebore and fit oversize piston and rings.
	Poor valve seatings	Grind in valves.
	Ignition timing incorrect	Check and adjust.
	Defective plug cap	Fit replacement.
	Dirty contact breaker points	Clean or renew.
Overheating	Accumulation of carbon on cylinder head	Decoke engine.
	Insufficient oil	Refill to specified level.
	Faulty oil pump and/or blocked oil passage	Strip and clean.
	Ignition timing too far retarded	Re-adjust.

61 Fault diagnosis: gearbox

Symptom	Cause	Remedy
Difficulty in engaging gears	Broken centre gear selector pawl or cam	Renew.
	Deformed gear selector	Repair or renew.
Machine jumps out of gear	Worn sliding gears on mainshaft and layshaft	Renew.
	Distorted or worn gear selector fork	Repair or renew.
	Weak gearchange drum stop spring	Renew spring
Gearchange lever fails to return normal position	Broken or displaced gearchange return spring	Renew or repair
Kickstart lever fails to return to normal position	Broken kickstart return spring	Renew spring.

Chapter 2 Clutch

Contents

General description ... 1	Clutch: reassembly ... 5
Clutch assembly: dismantling ... 2	Clutch: adjustment ... 6
Clutch: examination and renovation ... 3	Clutch: correct operation ... 7
Clutch operating mechanism: examination and renovation ... 4	Fault diagnosis: clutch ... 8

Specifications

Clutch springs
Number ... Four
Free length ... 19.6 mm (C50 model)
21.4 mm (C70 model)
27.0 mm (C90 model)
Minimum length ... 18.2 mm (C50 model)
20.4 mm (C70 model)
26.0 mm (C90 model)

Inserted clutch plates
Number ... 2 or 3, depending on model
Thickness ... 3.5 mm (C50 and C70 models)
2.9 mm (C90 model)
Minimum thickness ... 3.1 mm (C50 and C70 models)
2.4 mm (C90 model)

1 General description

The clutch is of the multi-plate type having two or three plain plates and two or three inserted plates depending on the model. The clutch is fully automatic in operation and interconnected with the gear change pedal so that it disengages and re-engages in the correct sequence.

2 Clutch assembly: dismantling

The clutch assembly complete is removed by following the procedure detailed in Chapter 1. Sections 12 and 13. When removed, the clutch can be broken down into its component parts as follows:

1 With the drive side (back) facing upwards, prise out the

2.1a Remove the large circlip ...

2.1b ... lift out the clutch centre ...

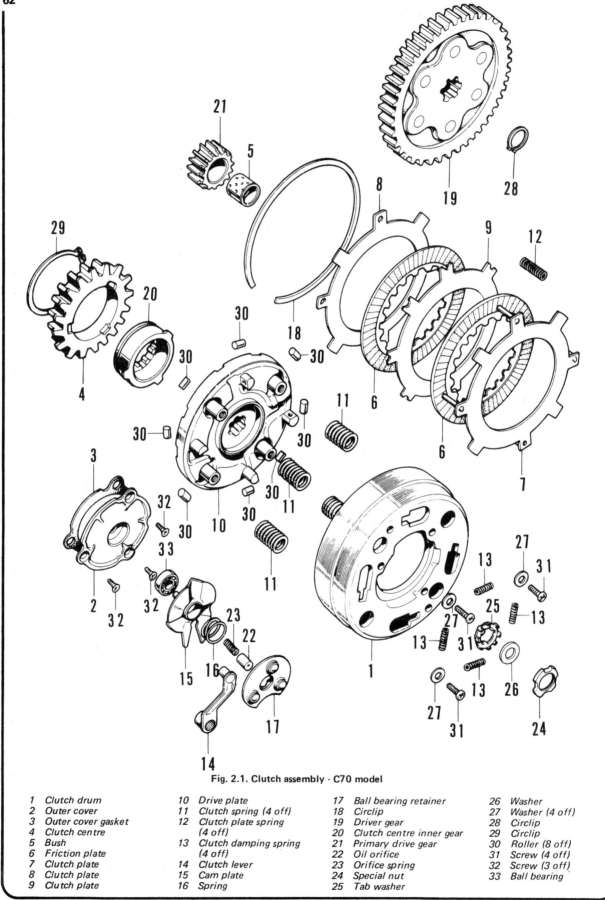

Fig. 2.1. Clutch assembly - C70 model

1 Clutch drum	10 Drive plate	17 Ball bearing retainer	26 Washer
2 Outer cover	11 Clutch spring (4 off)	18 Circlip	27 Washer (4 off)
3 Outer cover gasket	12 Clutch plate spring (4 off)	19 Driver gear	28 Circlip
4 Clutch centre	13 Clutch damping spring (4 off)	20 Clutch centre inner gear	29 Circlip
5 Bush		21 Primary drive gear	30 Roller (8 off)
6 Friction plate	14 Clutch lever	22 Oil orifice	31 Screw (4 off)
7 Clutch plate	15 Cam plate	23 Orifice spring	32 Screw (3 off)
8 Clutch plate	16 Spring	24 Special nut	33 Ball bearing
9 Clutch plate		25 Tab washer	

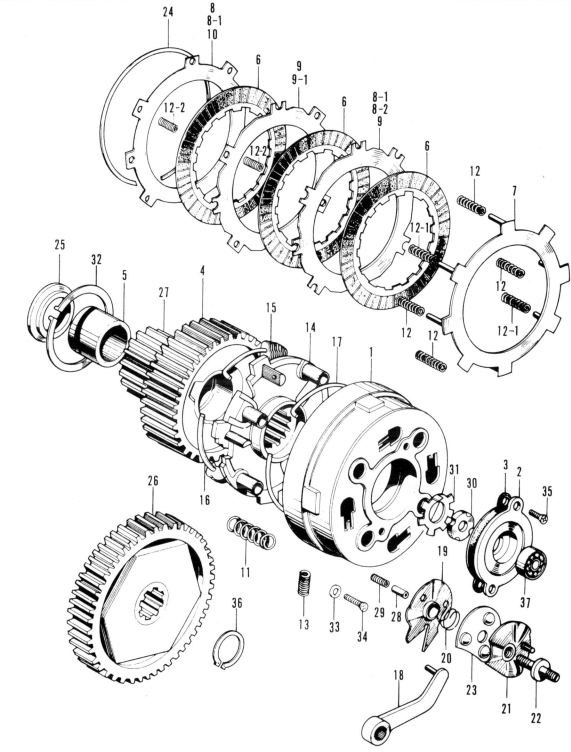

Fig. 2.2. Clutch assembly - C90 model

1 Clutch drum	11 Clutch spring (4 off)	19 Cam plate	29 Orifice spring
2 Outer cover	12 Clutch plate spring (6 off)	20 Spring	30 Special nut
3 Outer cover gasket		21 Adjustable cam plate	31 Tab washer
4 Clutch centre	13 Clutch damping spring (4 off)	22 Adjuster	32 Circlip
5 Bush		23 Ball bearing retainer	33 Washer (4 off)
6 Friction plate	14 Drive plate	24 Circlip	34 Screw (4 off)
7 Clutch plate	15 Bob-weight (24 off)	25 Spacer	35 Screw (2 off)
8 Clutch plate	16 Bob-weight clip	26 Drivegear	36 Circlip
9 Clutch plate	17 Bob-weight stop ring	27 Primary drive gear	37 Ball bearing
10 Clutch plate	18 Clutch lever	28 Oil orifice	

Chapter 2: Clutch

2.1c ... and remove the clutch plates

2.2 Remove the eight rollers which act as centrifugal weights

2.3 Remove the four screws to release the drive plate and springs

2.4 The circlip retains the clutch centre

3.4 Examine the clutch plates carefully

mm circlip from the rear of the clutch body and lift out the clutch centre assembly, complete with the clutch plates. The clutch plates will lift off the centre but care should be taken to avoid losing the four small plate separation springs that are located on the pins of the first clutch plate.

2 Remove the eight hardened steel rollers or on the C90 model the four shaped weights and clip.

3 Invert the clutch body and remove the four crosshead screws from the front face, unscrewing each a little at a time. This will release the drive plate, the four small damper springs and the four main clutch springs.

4 Removal of the retaining circlip from the clutch centre will permit the clutch drive gear to be separated.

3 Clutch: examination and renovation

1 Check the condition of the clutch drive to ensure none of the teeth are chipped, broken or badly worn.

2 Give the plain and the inserted clutch plates a wash with a paraffin/petrol mix and check that they are not buckled or distorted. Remove all traces of clutch insert debris, otherwise a gradual build-up will affect clutch action.

Chapter 2: Clutch

3 Visual inspection will show whether the tongues of the clutch plates have become burred and whether indentations have formed in the slots with which they engage. Burrs should be removed with a file, which can also be used to dress the slots, provided the depth of the indentations is not great.

4 Check the thickness of the friction linings in the inserted plates, referring to the Specifications section of this Chapter for the serviceable limits. If the linings have worn to, or below these limits, the plates should be renewed. Worn linings promote clutch slip.

5 Check also the free length of the clutch springs. The recommended serviceable limits are also in the Specification section. Do not attempt to stretch the springs if they have compressed. They must be renewed when they reach the serviceable limit, as a complete set.

6 Check the condition of the roller thrust bearing in the clutch outer plate.

4 Clutch operating mechanism: examination and renovation

The automatic clutch fitted to these models is designed so that as the engine speed increases, eight hardened steel rollers increase their pressure on the clutch plates through being thrown outwards along their respective tapered tracks by centrifugal force. Four small diameter compression springs assist the clutch plates to free, and four large diameter compression springs supply additional pressure when the rollers reach the end of their tracks.

A quick acting three-start thread mechanism is incorporated in an extension of the drive gear to apply pressure when the kickstart is operated, or when the machine is on the over-run.

The clutch is completely disengaged each time the gear operating pedal is moved, through a direct linkage between the gear change lever spindle and the clutch withdrawal mechanism.

1 Check the condition of the roller ramps in the clutch drive plate and the roller contact area. Excessive wear in these areas is often the cause of engine stalling, fierce clutch engagement and difficulty in gear changing. Replace the worn parts.

2 It is rarely necessary to replace the eight rollers or the clutch housing, unless the rollers show evidence of wear and the clutch housing has roller indentations. This type of wear is caused by poor gear changing, usually by releasing the gear pedal too fast when moving away from a standstill or changing gear.

3 The C90 model uses a different mode of operation involving the use of four weights. The earliest problem of wear is therefore obviated even though the clutch operates on the centrifugal principle.

5 Clutch: reassembly

1 Reassemble the clutch components by following the dismantling procedure in reverse.

2 The built-up clutch is then replaced on the splined end of the crankshaft, following the engine reassembly procedure given in Chapter 1, Section 50. A torque setting of 27.5 - 32.5 lbs ft is recommended for the sleeve nut that retains the clutch in position. Make sure the tab washer is bent over, to lock the sleeve nut in position.

3 Replace the oil filters and cover reassembly as described in Chapter 1, Section 51, ensuring that the clutch is adjusted according to the details given in the next section of this Chapter.

6 Clutch: adjustment

1 Clutch adjustment is provided by means of an adjustable screw and locknut located in the centre of the clutch cover. Slacken off the 10 mm locknut and turn the adjusting screw firstly in the clockwise direction, to ensure there is no end pressure on the clutch pushrod.

2 Turn the adjusting screw anti-clockwise until pressure can be felt on the end. Turn back (clockwise) for approximately 1/8th of a turn, and tighten the locknut, making sure the screw does not turn. Clutch adjustment should now be correct.

7 Clutch: correct operation

1 As the special starting mechanism operates when starting the engine, clutch slip can only be detected when the machine is being ridden, by the fact that the engine speed will increase with no increase in road speed.

2 Clutch drag is characterised by the engine having a tendency to stall or the machine starting to move forward, when first gear is engaged, with the engine running at tickover speed.

3 Refer to the fault diagnosis chart, Section 8, for possible causes for the above symptom.

4 Note that a fast tickover speed will cause the machine to snatch when first gear is engaged during the pullaway from a standstill.

3.6 Check the condition of the ball bearing

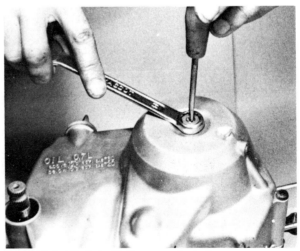

6.1 Clutch is adjusted with the cover attached

8 Fault diagnosis: clutch

Symptom	Cause	Remedy
Clutch slips	Incorrect adjustment	Re-adjust.
	Weak clutch springs	Renew set of four.
	Worn or distorted pressure plate	Renew.
	Distorted clutch plates	Renew.
	Worn friction plates	Renew.
Knocking noise from clutch	Loose clutch centre	Renew clutch.
Clutch does not fully disengage	Incorrect adjustment	Re-adjust
	Uneven clutch spring tension	Re-adjust.
	Distorted clutch plates	Renew.

Chapter 3 Fuel system and lubrication

Contents

General description ... 1	Carburettor: adjustments ... 11
Legshield: removal ... 2	Airfilter: location and cleaning ... 12
Petrol tank: removal and replacement ... 3	Exhaust system: cleaning ... 13
Petrol feed pipe: examination and renewal ... 4	Lubrication system ... 14
Petrol tap: removal and replacement ... 5	Oil filters and pressure relief valve: location and cleaning ... 15
Carburettor: general description ... 6	Trochoidal oil pump: description and location ... 16
Carburettor: removal ... 7	Trochoidal oil pump: removal, renovation and replacement ... 17
Carburettor: dismantling C50 and C70 models ... 8	
Carburettor: dismantling C90 model ... 9	Fault diagnosis: fuel system and lubrication ... 18
Carburettor: cleaning, examining and reassembling ... 10	

Specifications

Petrol tank

	C50	C70	C90
Capacity	3.0 litres (5.3 Imp pints)	4.5 litres (7.9 Imp pints)	5.5 litres (9.7 Imp pints)

Carburettor

	C50	C70	C90
Make	Keihin Seiki		
Type	DP13N13 (1000 - 110)	DP13N14AI (1000 - 112)	PW15HAI (1000 - 109)
Main jet	70	75	75
Slow running jet	35	35	40
Throttle slide	2.0	2.5	2.0
Needle	13239 - 3 stage	13243 - 3 stage	16332 - 2 stage
Slow running screw	1 - 1 1/4 turns out	1 3/8 - 1 5/8 turns out	3/8 - 1 1/8 turns out
Needle jet bore	3.0 mm	2.10 mm	2.60 mm

Oil pump

Outer rotor to pump body clearance	0.1 - 0.15 mm (0.004 - 0.006 inch)
Inner and outer rotor end clearance	0.02 - 0.07 mm (0.0008 - 0.0028 inch)
Rotor tip clearance	0.15 mm (0.006 inch)
replace if over	0.2 mm (0.008 inch)

1 General description

The fuel system comprises a fuel tank from which petrol is fed by gravity to the float chamber of the carburettor. The open frame layout necessitates the use of a specially-shaped petrol tank that is located immediately below the nose of the dualseat. On these models, the petrol tap is incorporated in the top of the carburettor float chamber.

All machines are fitted with a carburettor of Keihin manufacture, the model depending on the type of machine to which the carburettor is fitted. All carburettors have a manually-operated choke and employ a throttle slide and needle arrangement for controlling the petrol/air mixture administered to the engine.

2 Legshield: removal

1 The plastic legshield assembly encloses the engine unit, to enhance the clean appearance of the machine. There are suitable holes and blanking discs to enable the simplest of the adjusting tasks to be carried out without necessitating removal of the legshields, but whenever access is required for more complex tasks, the legshield assembly must be removed.
2 Place the machine on the centre stand and make sure it is standing firmly, on level ground.
3 Remove the side panels to obtain the toolkit and reveal the battery. Unscrew the fuse holder and remove the fuse, to isolate the battery, eliminating all risks of electrical mishap.
4 Remove the domed nut and the air cleaner lid.

Chapter 3: Fuel system and lubrication

3.5a Remove the four fixing bolts ...

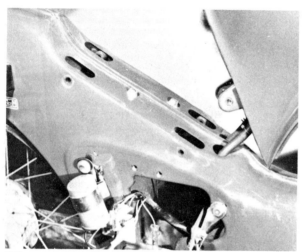

3.5b ... and release the petrol pipes

5.3 The dismantled petrol tap

5 Slacken the nuts holding the rear of the legshield and remove the clamping band, if fitted. Remove the four bolts holding the legshield and pull their spacers clear. The legshield assembly will now lift clear.

3 Petrol tank: removal and replacement

1 It is unlikely that there will be need to remove the petrol tank completely unless the machine has been laid up and rust has formed inside or it needs reconditioning. The engine/gear unit can be removed from the frame without having to detach the tank. The ignition coil is mounted inside the frame and the tank has to be removed first, to gain access.
2 As the petrol tap is an integral part of the carburettor, the petrol tank must be drained before it is removed.
3 Release the clips, pull both of the petrol feed pipes off the carburettor and let them drain into a suitable container.
4 Lift the dualseat and either remove it as described in Chapter 5 or fix it securely to avoid its falling forward at a crucial stage.
5 Remove the four bolts and washers holding the tank to the frame. Gently lift the tank, release the clips and pull off both of the petrol pipes. The tank will now lift clear. Leave the pipes threaded into the frame, to ease assembly.
6 To replace the tank, reverse the procedure described in the preceding paragraphs, ensuring that the petrol pipe with the red stripe is connected to the reserve pipes on the tank and carburettor.

4 Petrol feed pipes: examination and renewal

1 The petrol feed pipes are made of synthetic rubber and a check that they are not cracked or chafed, where they pass through the frame, should be made, as leaking petrol can cause a fire. Ensure that the wire retaining clips on each end are present, in good condition and properly located.
2 To renew the petrol feed pipes it is necessary to remove the petrol tank, as described in Section 3, and the battery as described in Chapter 7, to permit access to the clips and the inside of the frame. Ensure that the petrol pipe with the red stripe is connected to the reserve pipes on the tank and carburettor.

5 Petrol tap: removal and replacement

1 The petrol tap is a three position open - reserve - closed type, built as an integral part of the carburettor.
2 It can be dismantled without removing the carburettor but the petrol tank will have to be drained as described in Section 3.
3 Remove the two screws, the tap cover, the wave washer, the lever plate and the rubber sealing disc. Check for deterioration of the rubber disc and reassemble in the reverse order.
4 Refit the petrol pipes and refill the tank.

6 Carburettor: general description

Various types of Keihin carburettor are fitted to the Honda 50 cc - 90 cc models, the exact specifications depending on the designation of the model. Air is drawn into all the carburettors, via an air filter with a removable element. The conventional throttle slide and needle arrangement works in conjunction with the main jet, to control the amount of petrol/air mixture administered to the engine. There is also a slow running jet with an adjustable air screw, to control idling at low speeds, and a manually-operated choke, to aid cold starting.

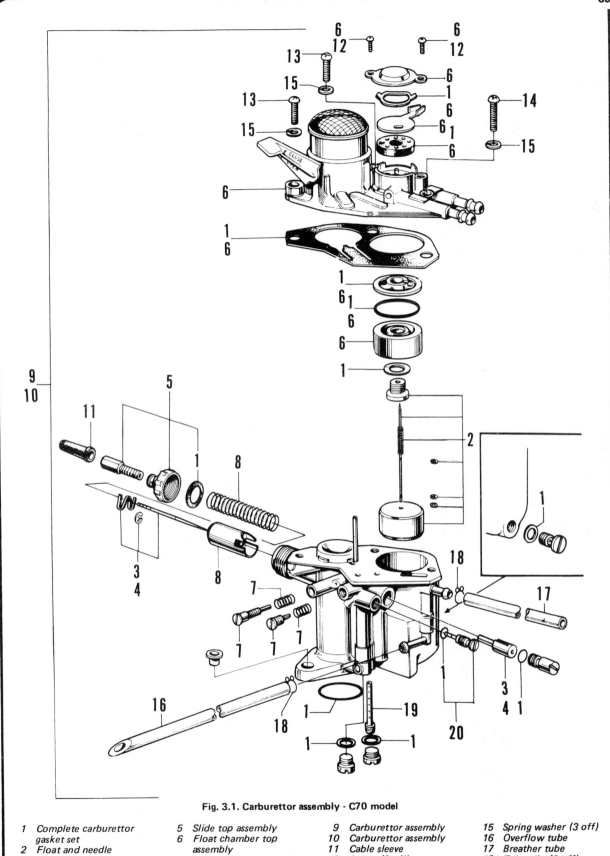

Fig. 3.1. Carburettor assembly - C70 model

1 Complete carburettor gasket set
2 Float and needle assembly
3 Needle and jet set
4 Needle and jet set
5 Slide top assembly
6 Float chamber top assembly
7 Slow running and throttle stop screws
8 Slide and return spring
9 Carburettor assembly
10 Carburettor assembly
11 Cable sleeve
12 Screw (2 off)
13 Screw (2 off)
14 Screw
15 Spring washer (3 off)
16 Overflow tube
17 Breather tube
18 Tube clip (2 off)
19 Main jet
20 Slow running jet

7 Carburettor: removal

Engine removed from the frame
1   If the engine has already been removed from the frame and, on the C90 model, the inlet tube removed, the carburettor will be attached to the control cable, the petrol pipes and to the synthetic rubber intake tube. Release the clip and pull the carburettor out of the rubber tube after the petrol tank has been drained as described in Section 3.

Engine still in the frame
2   If the engine is still in the frame the carburettor is sandwiched between the top of the engine and the main frame tube.
3   Drain the petrol tank as described in Section 3.
4   Disconnect the wires to the horn at the snap connectors. Release the spring clips at both ends of the rubber intake tube and pull the tube clear.
5   Remove the two nuts and washers and pull the carburettor off the engine or the inlet tube.
6   Make sure the 'O' ring in the carburettor flange is not lost.

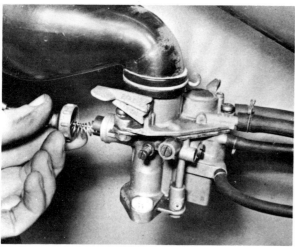
8.1 Unscrew the top and withdraw the slide

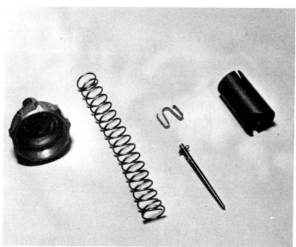
8.2 The dismantled slide assembly

8.3a Remove the screws and the float chamber top ...

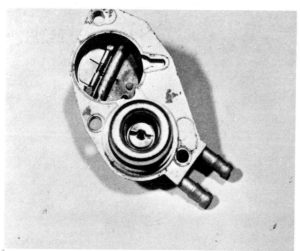
8.3b ... and unscrew the float valve seat ...

8.3c ... and remove the petrol filter

Chapter 3: Fuel system and lubrication

8 Carburettor: dismantling (C50 and C70 models)

1 At this stage, the carburettor is still attached to the control cable and to remove it completely the carburettor top has to be unscrewed and the slide and needle pulled out.
2 Compress the slide return spring and unhook the throttle cable. The slide, the needle with its spring clip, the w-shaped spring, the return spring and the carburettor top will then slide off the cable.
3 Remove the three screws holding the float chamber top in position and pull the top clear. Remove the float needle seat from the underside of the float chamber top. Lift the filter chamber, the sealing ring, and the petrol filter out of the float chamber top.
4 Lift the float and needle assembly out of the float chamber.
5 Remove the throttle stop screw and the slow running air screw from the side of the carburettor, taking care not to lose the small springs.
6 Remove the blanking plug, with its 'O' ring, and push out from the throttle slide chamber the needle jet. Remove the adjacent slow running jet.

8.4 The float assembly will lift out

8.5a Remove the throttle stop screw ...

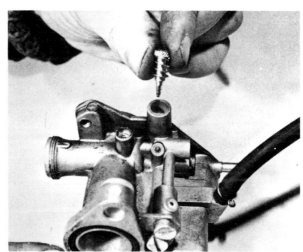
8.5b ... and the slow running mixture screw

8.6a Remove the sealing plug ...

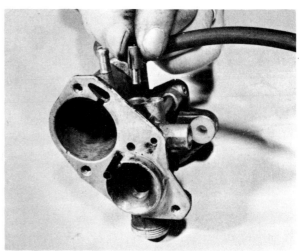
8.6b ... and the needle jet from one hole ...

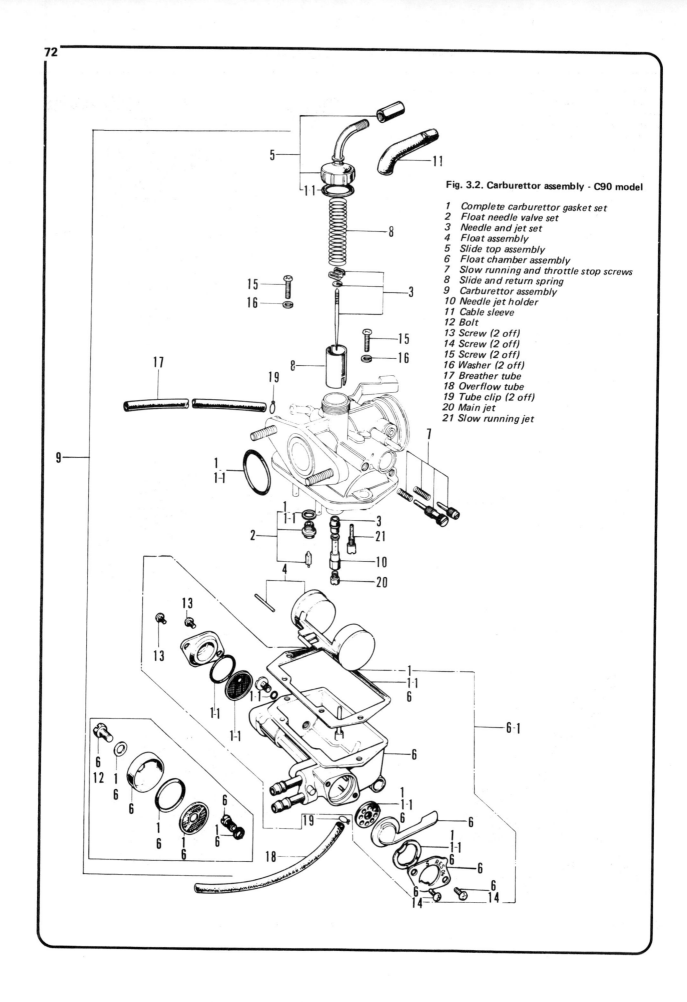

Fig. 3.2. Carburettor assembly - C90 model

1 Complete carburettor gasket set
2 Float needle valve set
3 Needle and jet set
4 Float assembly
5 Slide top assembly
6 Float chamber assembly
7 Slow running and throttle stop screws
8 Slide and return spring
9 Carburettor assembly
10 Needle jet holder
11 Cable sleeve
12 Bolt
13 Screw (2 off)
14 Screw (2 off)
15 Screw (2 off)
16 Washer (2 off)
17 Breather tube
18 Overflow tube
19 Tube clip (2 off)
20 Main jet
21 Slow running jet

Chapter 3: Fuel system and lubrication

7 Remove the two blanking plugs that are close to the mounting flange and remove the main jet, using a small screwdriver, from the carburettor body.
8 Remove the float chamber drain screw.
9 The choke flap is rivetted into position and cannot easily be removed.

9 Carburettor: dismantling (C90 model only)

1 At this stage, the carburettor is still attached to the control cable and to remove it completely, the carburettor top has to be unscrewed and the slide and needle pulled out.
2 Compress the slide return spring and unhook throttle cable. The slide, the needle with its spring clip, the w-shaped spring, the return spring and the carburettor top will then slide off the cable.
3 Remove the two screws holding the float chamber in position and pull the chamber clear. Remove the two screws on the end of the float chamber to release the blanking plate, sealing ring and petrol filter. Remove the float chamber drain screw.
4 Invert the carburettor body and push out the float pivot pin. This releases the float and the float needle. Carefully remove the float needle from the float.
5 Remove from the underside of the carburettor body the float needle seat, the slow running jet and from the centre the main jet, the jet holder and the needle jet.
6 Remove the throttle stop screw and the slow running air screw from the side of the carburettor, taking care not to loose the small springs.
7 The choke flap is rivetted into position and cannot easily be removed.

10 Carburettor: cleaning, examining and reassembling

1 Thoroughly clean all the parts paying particular attention to the internal passageways of the carburettor body, the bottom of the float chamber and any other places where sediment may collect.
2 Check that none of the springs are weak or broken.
3 Check for wear on the slide and carburettor body as air leaks round the slide can cause weak mixture problems.
4 Check for ridges on the conical sealing portion of the float needle.
5 Check the condition of the float and shake it to see if there is

8.6c ... and the slow running jet from the one adjacent

8.7a Remove the two sealing plugs ...

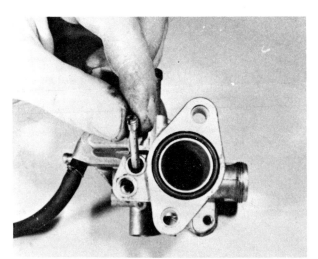

8.7b ... and the main jet

8.8 Remove the float chamber drain screw

any petrol inside. The float is non-repairable and should be renewed if damaged or punctured.

6 Check that all gaskets, sealing washers or rubber seals are in good condition. Preferably renew them when reassembling as leaking petrol can cause a fire.

7 When reassembling the carburettor, follow the dismantling instructions in reverse, ensuring that the needle clip is in its correct groove.

8 The various sizes of the jets, throttle slide and needle are predetermined by the manufacturer and should not require modification. Check with the Specifications list if there is any doubt about the values fitted.

11 Carburettor: adjustments

1 All adjustments should be made when the engine is at normal working temperature.

2 To adjust the slow running speed the throttle cable should be slackened to ensure that it is the throttle stop screw that is holding the slide and not the cable. Set the throttle stop screw so that the engine runs at a fast tick-over speed.

3 Screw in or out the air screw until the engine runs evenly, without hunting or misfiring. Reduce the engine speed by unscrewing the throttle stop and re-adjust the air screw, if necessary. Do not arrive at a setting where the engine ticks over too slowly, otherwise there is risk that it may stall when the throttle is closed, during normal running.

4 As a rough guide, the air screw should be positioned from one to one and a quarter complete turns out from the fully closed position.

5 The amount of throttle slide cutaway, size of main jet, size of needle jet and size of slow running jet are pre-determined by the manufacturer and should be correct for the model in which they are used. Check with the Specifications, page 67. The throttle needle position can be varied, by removing and replacing the needle clip. Under normal circumstances, the needle should be positioned in the second or third notch, measured from the top.

6 The slow running jet controls engine speed up to approximately 1/8th throttle and the degree of throttle slide cutaway from 1/8th to ¼ throttle. Thereafter the needle jet takes over, up to ¾ throttle, and main jet size controls the final ¾ to full throttle. These stages are only approximate; there is a certain amount of overlap.

7 Readjust the throttle cable to remove any excess play but leave a small amount of slack to avoid the engine speed increasing when the handlebars are turned.

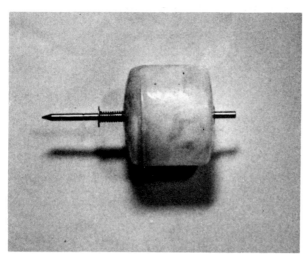

10.5 Check the float assembly for damage

12.1 Clean the air filter carefully as it is only paper

13.2a Remove the swinging arm nut to release the silencer ...

13.2b ... and the collets to release the exhaust pipe

Chapter 3: Fuel system and lubrication

12 Air filter: location and cleaning

1 The air filter is located on top of the main frame tube, immediately behind the steering head. It is clearly visible when the the legshield assembly has been removed.
2 To clean the air filter, remove the detachable element and tap it lightly to remove accumulated dust. Blow from the inside de with compressed air, or brush the exterior with a light brush. Remember the element is made from paper. If it is torn or damaged, fit a replacement.
3 Oil or water will reduce the efficiency of the filter element and may upset the carburation. Renew any suspect element.
4 It is advisable to replace the element at less than the recommended 6,000 miles if the machine is used in very dusty conditions. The usual signs of a filter element in need of replacement are reduced performance, misfiring and a tendency for the carburation to run rich.
5 On no account should the machine be run without the filter element in place because this will have an adverse effect on carburation.

13 Exhaust system: cleaning

1 Although the exhaust system on a four-stroke does not require such frequent attention as that of the two-stroke, it is nevertheless advisable to inspect the complete system from time to time in order to ensure a build-up of carbon does not cause back pressure. If an engine is nearing the stage where a rebore is necessary, it is advisable to check the exhaust system more frequently. The oily nature of the exhaust gases will cause a more rapid build-up of sludge.
2 The complete exhaust system is removed from the machine by detaching the swinging arm nut, the two nuts and flange at the exhaust port, and pulling the exhaust clear. The two half collets will fall clear from the exhaust pipe but the exhaust gasket will need to be prised out of the cylinder head and consequently will need renewing. If this joint is not an airtight seal, the engine will tend to backfire on the over-run.
3 A 10 mm bolt in the extreme end of the silencer retains the detachable baffle assembly in position. If this bolt is withdrawn, the baffle tube can be pulled clear of the silencer body, for cleaning.
4 Tap the baffle to remove loose carbon and work with a wire brush, if necessary. If there is a heavy build-up of carbon or oily sludge, it may be necessary to use a blow lamp to burn out these deposits.
5 The exhaust pipe and silencer are one unit and if a large amount of carbon has built up inside it is necessary to fill the silencer with a solution of caustic soda after blocking up one end. If possible, leave the caustic soda solution within the silencer overnight, before draining off and washing out thoroughly with water.
6 Caustic soda is highly corrosive and every care should be taken when mixing and handling the solution. Keep the solution away from the skin and more particularly the eyes. The wearing of rubber gloves is advised whilst the solution is being mixed and used.
7 The solution is prepared by adding 3 lbs of caustic soda to 1 gallon of COLD water, whilst stirring. Add the caustic soda a little at a time and NEVER add the water to the chemical. The solution will become hot during the mixing process, which is why cold water must be used.
8 Make sure the used caustic soda solution is disposed of safely, preferably by diluting with a large amount of water. Do not allow the solution to come into contact with aluminium castings because it will react violently with this metal.
9 To reassemble the exhaust system reverse the dismantling procedure, ensuring that the baffle assembly retaining bolt is fully tightened.
10 Do not run the machine without the baffle tube in position. Although the changed engine note may give the illusion of greater speed, the net effect will be a marked drop in performance as a result of changes in carburation. There is also risk of prosecution as a result of the excessive noise.

14 Lubrication system

1 Oil is picked up from the oil compartment in the crankcase by the oil pump, via an oil filter screen which filters out any impurities that may otherwise damage the pump itself. The pump delivers oil, under pressure, to the right-hand crankcase where it follows the routes listed below:

a) *The oil passes through a drilling in the clutch cover, through the pressure release orifice in the centre of the clutch, through the centrifugal oil filter, and into the crankshaft, to lubricate the big end and main bearings.*
b) *The oil passes up the side of one of the holding down studs, through the cylinder barrel and into the cylinder head, where a side cover distributes the oil to the rocker pins and the camshaft. Some of the oil lubricates the camchain on its return, although the C50 and C70 models have an oil return passageway.*
c) *On the C90 model only, there are additional drillings within the crankcases to feed oil to the mainshaft and layshaft plain bearings.*

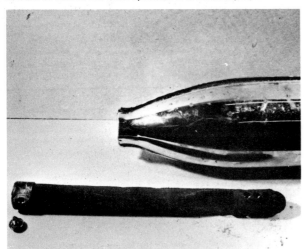

13.3 The baffle tube is removable for cleaning

15.3 The gauze oil filter fits in a slot in the crankcase

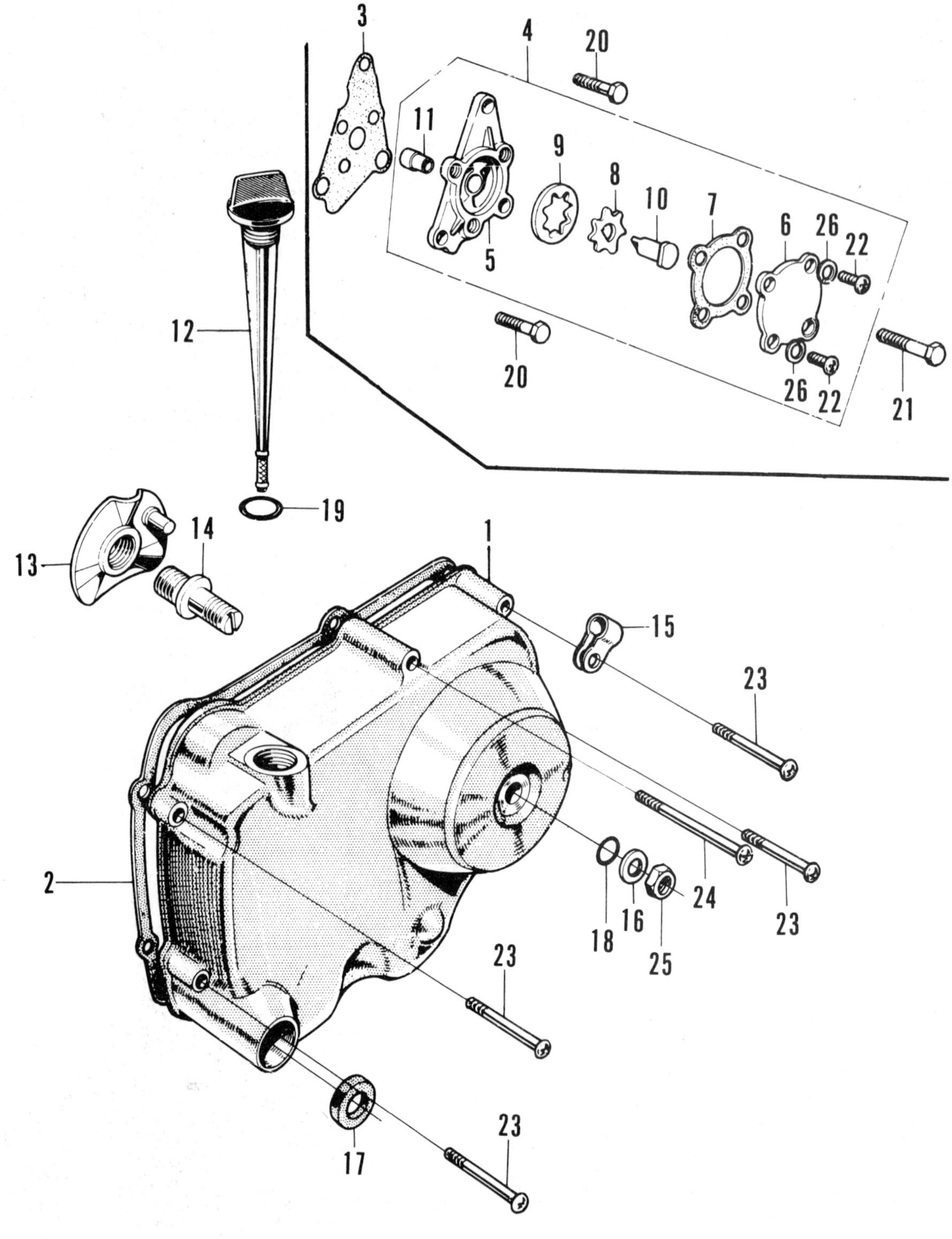

Fig. 3.3. Clutch cover and oil pump - C70 model

1 Clutch cover
2 Clutch cover gasket
3 Oil pump gasket
4 Oil pump assembly
5 Oil pump body
6 Oil pump cover
7 Cover gasket
8 Inner rotor
9 Outer rotor
10 Driveshaft
11 Dowel
12 Dipstick
13 Adjustable camplate
14 Adjuster
15 Clip
16 Washer
17 Oil seal
18 O-ring
19 O-ring
20 Bolt (2 off)
21 Bolt
22 Screw (3 off)
23 Screw (6 off)
24 Screw (2 off)
25 Nut
26 Washer (3 off)

Chapter 3: Fuel system and lubrication

2 The remainder of the engine components are lubricated by splash from the oil content of the sump.

15 Oil filters and pressure relief valve: location and cleaning

1 As explained in the previous Section, there are two filters in the lubrication system, a square section gauze filter screen that slots into a cavity in the right-hand crankcase, and a filter of the centrifugal type that is attached to the outer face of the clutch.
2 Both can be removed for cleaning when the procedure given in Chapter 1 Section 12 is followed.
3 The gauze filter should be cleaned by immersing it in petrol and if necessary, brushing it with a soft haired brush to remove any impurities or foreign matter. Allow it to dry before replacement. If for any reason the gauze is damaged, the complete filter must be renewed.
4 The centrifugal filter should be washed out with petrol and any impurities or foreign matter removed in similar fashion. Dry the assembly with clean rag, prior to reassembly. Before replacing the end cover and tightening the screws, check the condition of the sealing gasket.
5 When using petrol for washing purposes, take extreme care as petrol vapour is highly inflammable. Cleaning should preferably be accomplished in the open air or in well-ventilated surroundings away from any naked flames.
6 In the centre of the clutch release mechanism there is a spring loaded orifice that acts as a pressure relief valve. Ensure that the hole in the orifice is clear, that the spring is in good condition and that the orifice is free to move within the release mechanism.
7 For the reassembly sequence, refer to Chapter 1 Section 51.

16 Trochoidal oil pump: description and location

1 The trochoid oil pump is located behind the clutch, where it is retained to the right-hand crankcase by three bolts. It is extremely unlikely that the pump will require attention under normal circumstances and should not be dismantled unnecessarily.
2 The pump comprises an inner and an outer rotor. The pumping action is provided by the differences in the shape and number of teeth between the inner and the outer rotors.

15.4 The clutch outer plate forms the centrifugal oil filter

15.6 Check the pressure relief valve

17.2 Remove the cover to check oil pump operation

17.5 Remove the pump to renew any worn parts

17 Trochoidal oil pump: removal, renovation and replacement

1 If the pump is suspected in the event of a lubrication failure, it can be dismantled after it has been detached from the engine unit by referring to Sections 12, 13 and 23 of Chapter 1.
2 Remove the three screws and the rotor cover plate to gain access to the inner and outer rotors.
3 Clearances between the various internal components can be checked with the wear limits given in the Specifications Section of this Chapter.
4 To reassemble the oil pump, fit the outer rotor, the inner rotor and feed the drive shaft into position rotating the rotors if necessary to fully engage the shaft. Stick the cover gasket in position, using a very thin film of Golden Hermatite, and refit the cover and screws.
5 Refit the oil pump as described in Chapter 1 Section 40 and complete the engine reassembly as described in Chapter 1 Sections 50 and 51.

18 Fault diagnosis: fuel system and lubrication

Symptom	Cause	Remedy
Excessive fuel consumption	Air cleaner choked or restricted	Clean or renew.
	Fuel leaking from carburettor. Float sticking	Check all unions and gaskets. Float needle seat needs cleaning.
	Badly worn or distorted carburettor	Renew.
	Jet needle setting too high	Adjust to figure given in Specifications.
	Main jet too large or loose	Fit correct jet or tighten if necessary.
	Carburettor flooding	Check float valve and renew if worn.
Idling speed too high	Throttle stop screw in too far.	Adjust screw.
	Carburettor top loose	Tighten top.
	Pilot jet incorrectly adjusted	Refer to relevant paragraph in this Chapter.
	Throttle cable sticking	Disconnect and lubricate or replace.
Engine dies after running a short while	Blocked air hole in filler cap	Clean.
	Dirt or water in carburettor	Remove and clean out.
General lack of performance	Weak mixture; float needle stuck in seat	Remove float chamber or float and clean.
	Air leak at carburettor joint	Check joint to eliminate leakage, and fit new O-ring.
Engine does not respond to throttle	Throttle cable sticking	See above.
	Petrol octane rating too low	Use higher grade (star rating) petrol.
Engine runs hot and is noisy	Lubrication failure	Stop engine immediately and investigate cause. Slacken cylinder head nut to check oil circulation. Do not restart until cause is found and rectified.

Chapter 4 Ignition system

Contents

General description ... 1	replacement ... 6
Legshield: removal ... 2	Condenser: removal and replacement ... 7
Flywheel generators: checking output ... 3	Ignition timing: checking and re-setting ... 8
Ignition coil: checking, removal and replacement ... 4	Automatic advance unit: location and checking action ... 9
Contact breaker: adjustment ... 5	Sparking plug: checking and resetting gap ... 10
Contact breaker assembly: removal, renovation and	Fault diagnosis: ignition system ... 11

Specifications

	C50	C70	C90
Generator			
Make	Hitachi Seisakusho	Mitsubishi Denki or Nippon Denso	Kokusan Denki
Type	F120	FAZ or 37000-026-0	EG26
Coil			
Make	Hitachi Seisakusho	Hitachi Seisakusho	Kokusan Denki
Type	CM61 - 08	CM61 - 08	ST 78
Spark plug			
Make	NGK or Champion	NGK or Champion	NGK or Champion
Type	C - 7HS Z - 8 or Z - 10	C - 7HS Z - 8 or Z - 10	D - 6HN or P7
Gap	0.6 - 0.7 mm (0.024 - 0.028 inch)	0.6 - 0.7 mm (0.024 - 0.028 inch)	0.6 - 0.7 mm (0.024 - 0.028 inch)
Contact breaker gap	0.3 - 0.4 mm (0.012 - 0.016 inch)	0.3 - 0.4 mm (0.012 - 0.016 inch)	0.3 - 0.4 mm (0.012 - 0.016 inch)

1 General description

The system used for producing the spark which is necessary to ignite the petrol/air mixture in the combustion chamber differs slightly between that used for the C50 and C70 models and the C90 model.

In the C50 and C70 system, the flywheel generator produces the electrical power which is fed directly to the ignition coil, mounted inside the frame. The condenser and contact breaker assembly are mounted inside the flywheel generator, where, with the help of an automatic advance and retard mechanism, they determine the exact moment at which the spark will occur. The ignition switch shorts out the ignition system when it is switched off.

In the C90 system, the battery produces the electrical power, which is fed through a fuse and the ignition switch to the ignition coil, mounted inside the frame. The condenser is mounted on the end of the ignition coil. The contact breaker assembly and automatic advance and retard mechanism are mounted on the end of the camshaft, on the left-hand side of the cylinder head. The generator will produce sufficient power for starting the engine if the battery is flat.

When the contact breaker points separate, the electrical circuit is interrupted and a high tension voltage is developed across the points of the spark plug, which jumps the air gap and ignites the mixture.

2 Legshield: removal

1 The plastic legshield assembly encloses the engine unit to enhance the clean appearance of the machine. There are suitable holes and blanking discs to enable the simplest of the adjusting tasks to be carried out without removing the legshield, but whenever access is required for more complex tasks the legshield assembly must be removed.
2 Place the machine on the centre stand and make sure it is standing firmly on level ground.
3 Remove the side panels to obtain the toolkit and reveal the battery. Unscrew the fuse holder and remove the fuse to isolate the battery, eliminating all risks of electrical mishap.
4 Remove the domed nut and the air cleaner lid.
5 Slacken the nuts holding the rear of the legshield assembly and remove the clamping band if fitted. Remove the four bolts holding the legshields and pull their spacers clear. The legshields will now lift clear.

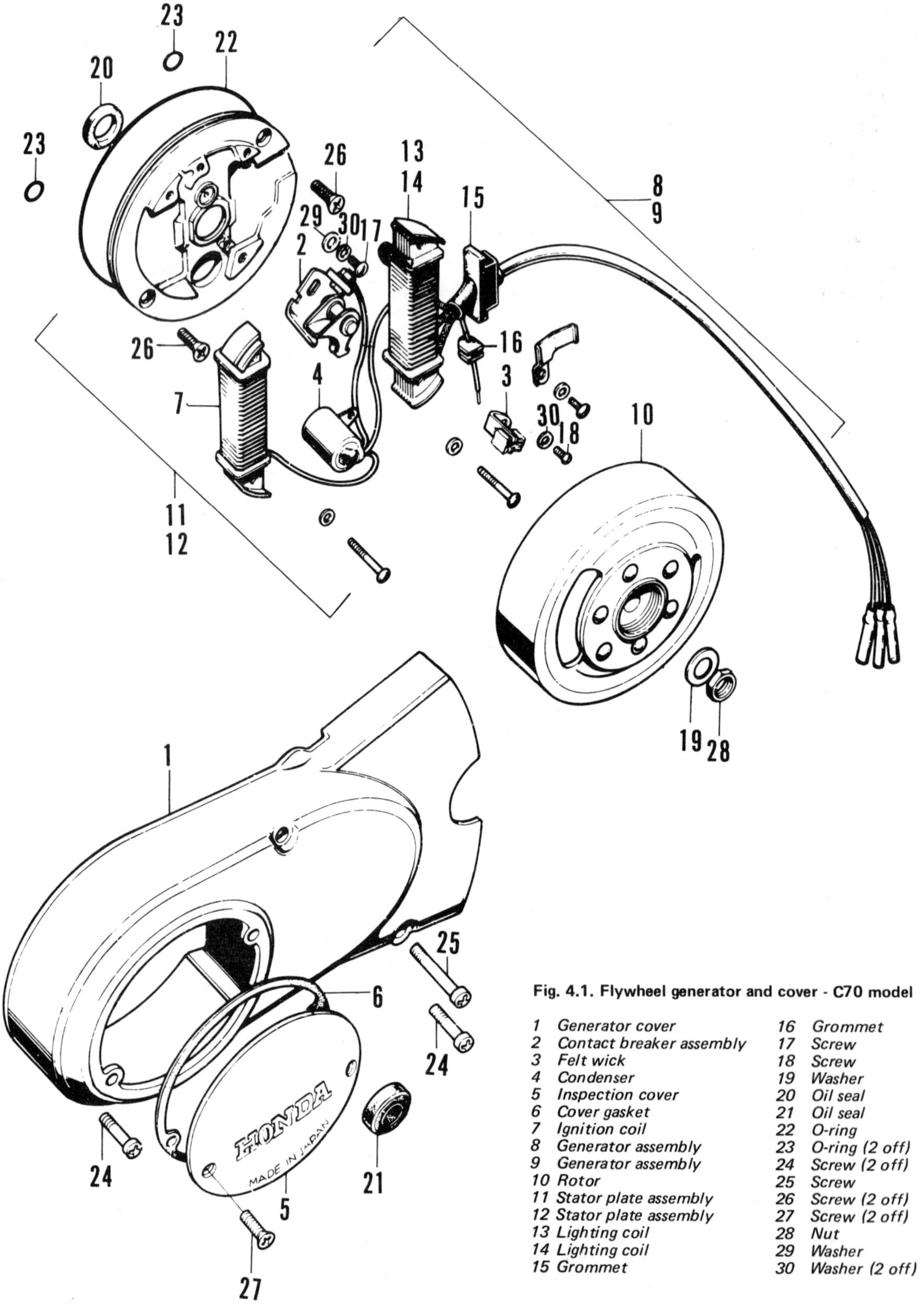

Fig. 4.1. Flywheel generator and cover - C70 model

1 Generator cover
2 Contact breaker assembly
3 Felt wick
4 Condenser
5 Inspection cover
6 Cover gasket
7 Ignition coil
8 Generator assembly
9 Generator assembly
10 Rotor
11 Stator plate assembly
12 Stator plate assembly
13 Lighting coil
14 Lighting coil
15 Grommet
16 Grommet
17 Screw
18 Screw
19 Washer
20 Oil seal
21 Oil seal
22 O-ring
23 O-ring (2 off)
24 Screw (2 off)
25 Screw
26 Screw (2 off)
27 Screw (2 off)
28 Nut
29 Washer
30 Washer (2 off)

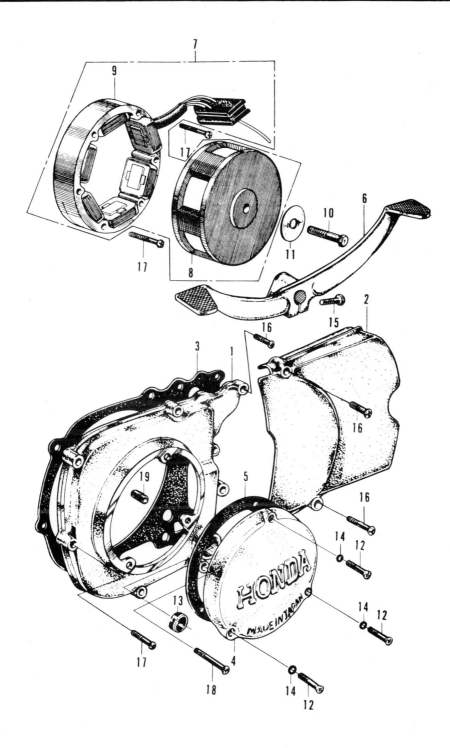

Fig. 4.2. Alternator and covers - C90 model

1 Alternator cover
2 Sprocket cover
3 Alternator cover gasket
4 Inspection cover
5 Cover gasket
6 Gearchange lever
7 Alternator assembly
8 Rotor
9 Stator assembly
10 Bolt
11 Washer
12 Screw (3 off)
13 Oil seal
14 O-ring (3 off)
15 Bolt
16 Screw (3 off)
17 Screw (9 off)
18 Screw
19 Dowel

Chapter 4: Ignition system

3 Flywheel generators: checking outputs

The output from either of the two types of generator used can be checked only with specialised test equipment of the multi-meter type. It is unlikely that the average owner/rider will have access to this equipment or instruction in its use. In consequence, if the performance of a generator is suspect, it should be checked by a Honda agent or an auto-electrical expert.

4 Ignition coil: checking, removal and replacement

1 The ignition coil is a sealed unit, designed to give long service. If a weak spark and difficult starting cause its performance to be suspect, it should be tested by a Honda agent or an auto-electrical expert. A faulty coil must be renewed. It is not practicable to effect a repair.
2 To gain access to the coil, the petrol tank must first be removed as described in Chapter 3, Section 3. The battery and its holder will also have to be removed, as described in Chapter 7, Section 3 to gain access to the inside of the frame where the coil is mounted and retained by two nuts. Remove the nuts and carefully pull the coil part way out so that the wires can be disconnected at the snap connectors. Pull the plug cap off the plug, unscrew it from the plug lead and feed the plug lead through the clip and back into the frame above the engine, to allow the ignition coil to pull clear.
3 Reassembly is the reverse of the removal procedure.

5 Contact breaker: adjustment

1 The C50 and C70 models use an ignition system commonly referred to as an 'energy transfer' system. The subtle advantage of this system is that the contact breaker adjustment is used to adjust the ignition timing. The resulting maximum contact breaker gap should be between 0.3 mm and 0.4 mm (0.012 in and 0.016 in). If it is outside these limits, the contact breaker assembly should be renewed, since it is worn out. The ignition timing section describes the method of adjustment. If it should be necessary to remove the contact breaker assembly for further attention, or renewal, it will be necessary to withdraw the flywheel generator from the crankshaft, following the procedure described in Chapter 1, Section 7.
2 The C90 model uses a conventional ignition system in which the contact breaker gap is adjusted first and the backplate adjusted to obtain the correct ignition timing.

3 Remove the two screws and the contact breaker cover on the left-hand side of the cylinder head.
4 Rotate the engine until the contact breaker is in its fully open position. Examine the faces of the contacts. If they are pitted or burnt it will be necessary to remove them for further attention, as described in Section 6 of this Chapter.
5 Check the contact breaker gap to see if it is between 0.3 mm and 0.4 mm (0.012 and 0.016 in). To adjust the contact breaker gap, slacken the two screws that hold the contact breaker assembly and using a small screwdriver in the slot provided, ease the assembly to the correct position. Tighten the screws and recheck the gap to ensure that the assembly has not moved.
6 It is always advisable to check the ignition timing, especially if the contact breaker gap has been reset. It will almost certainly require readjustment in this latter case.
7 Ensure that the sealing gasket is either renewed or is in good condition, before refitting the contact breaker cover and screws.

6 Contact breaker assembly: removal, renovation and replacement

1 If the contact breaker points are burned, pitted or badly worn, they should be removed for dressing. If it is necessary to remove a substantial amount of material before the faces can be restored, the points should be renewed.
2 To remove the contact breaker assembly, access must be gained as described in the preceding Section. Slacken and remove the nut at the end of the moving contact return spring. Remove the spring and plain washer and detach the spring. Note that an insulating washer is located beneath the spring, to prevent the electrical current from being earthed.
3 Remove the spring clip from the moving contact pivot and the insulating washer. Withdraw the moving contact, which is integral with the fibre rocker arm.
4 Remove the screws that retain the fixed contact plate and withdraw the plate complete with contact.
5 The points should be dressed with an oilstone or fine emery cloth. Keep them absolutely square during the dressing operation, otherwise they will make angular contact when they are replaced and will quickly burn away.
6 Replace the contacts by reversing the dismantling procedure. Take particular care to replace the insulating washers in the correct sequence, otherwise the points will be isolated electrically and the ignition system will not function.

4.2 The ignition coil is mounted inside the frame

6.3 Remove the screw and release the wire to withdraw the contact breaker assembly

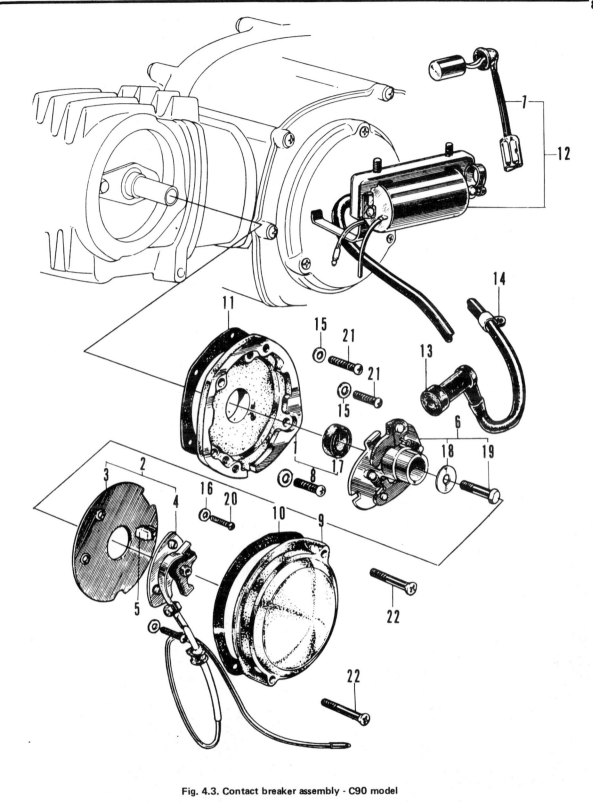

Fig. 4.3. Contact breaker assembly - C90 model

1 Contact breaker base	6 Automatic advance retard assembly	11 Base gasket	17 Oil seal
2 Contact breaker assembly		12 Coil assembly	18 Washer
3 Backplate	7 Condenser	13 Suppressor cap	19 Bolt
4 Contact breaker assembly	8 Base assembly	14 Clip	20 Screw (2 off)
5 Felt wick	9 Cover	15 Washer (3 off)	21 Screw (3 off)
	10 Cover gasket	16 Washer (2 off)	22 Screw (2 off)

Chapter 4: Ignition system

7.5 The condenser is mounted on the stator plate

8.2 When the 'F' mark lines up with the crankcase mark the points should be about to open

9.1 Check the action of the automatic advance/retard mechanism

7 Condenser: removal and replacement

1 A condenser is included in the contact breaker circuitry to prevent arcing across the contact breaker points as they separate. It is connected in parallel with the points and if a fault develops, ignition failure will occur.
2 If the engine is difficult to start or if misfiring occurs, it is possible that the condenser has failed. To check, separate the contact breaker points by hand whilst the ignition is switched on. If a spark occurs across the points and they have a blackened or burnt appearance, the condenser can be regarded as unserviceable.
3 It is not possible to check the condenser without the necessary test equipment. It is best to fit a replacement condenser and observe the effect on engine performance, especially in view of its low cost.
4 To remove the condenser on the C90 model, follow the procedure described in Section 4, for removing the ignition coil, as the condenser is clamped to the end of this coil. Slacken the clamp and pull the condenser clear.
5 To remove the condenser on the C50 and C70 models, follow the procedure described in Chapter 1, Section 7, for removing the flywheel generator, as the condenser is mounted on the stator plate. Unsolder the wires on the condenser, remove the fixing screw and pull the condenser clear.
6 Reassemble by reversing the dismantling procedure. Take care not to overheat the condenser when resoldering the wires into position as the insulation is very easily damaged by heat.

8 Ignition timing: checking and re-setting

1 To check the ignition timing, remove the generator inspection cover or, in the case of the C90 model, the contact breaker cover on the left-hand side of the cylinder head.
2 If the ignition timing is correct, the contact breaker points will be about to separate when the 'F' line scribed on the rotor of the flywheel coincides exactly with an arrow or an indentation or similar scribe mark on the left-hand crankcase or cover.
3 On the C50 and C70 models having the 'energy transfer' system, the ignition timing is varied by adjusting the contact breaker gap. Slacken the screw holding the contact breaker assembly, and using a small screwdriver in the slot provided, ease the assembly to the correct position. Tighten the screw and recheck the ignition timing. Check that the contact breaker gap is between 0.3 mm and 0.4 mm (0.012 in and 0.016 in), renewing the assembly, if outside these limits.
4 On the C90 models, which have a more conventional system, the ignition timing is adjusted by moving the contact breaker backplate. Before checking the ignition timing, always make sure the contact breaker gap is correct first. The backplate holding the complete contact breaker assembly is slotted, to permit a limited range of adjustment. If the two crosshead screws are slackened a little, the plate can be turned until the points commence to separate, and then locked in this position by tightening the screws.
5 After checking the timing, rotate the engine and check again before replacing the covers. The accuracy of the ignition timing is critical in terms of both engine performance and petrol consumption. Even a small error in setting will have a noticeable effect.

9 Automatic advance unit: location and checking action

1 Fixed ignition timing is of little advantage as the engine speed increases and provision is made to advance the timing by centrifugal means, using a balance weight assembly located behind the contact breaker assembly or within the rotor of the flywheel magneto generator. A check is not needed unless the action of the unit is in doubt.
2 To check the action of the unit it is first necessary to withdraw the contact breaker assembly complete from the C90

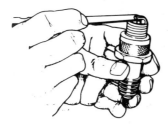
Checking plug gap with feeler gauges

Altering the plug gap. Note use of correct tool

Fig. 4.4a. Spark plug maintenance

White deposits and damaged porcelain insulation indicating overheating

Broken porcelain insulation due to bent central electrode

Electrodes burnt away due to wrong heat value or chronic pre-ignition (pinking)

Excessive black deposits caused by over-rich mixture or wrong heat value

Mild white deposits and electrode burnt indicating too weak a fuel mixture

Plug in sound condition with light greyish brown deposits

Fig. 4.4b. Spark plug electrode conditions

model or withdraw the rotor of the flywheel magneto generator. Refer to Chapter 1.7 for the dismantling procedure.

3 The counterweights of the automatic advance unit should return to their normal position with smooth action when they are spread apart with the fingers and released. A visual inspection will show signs of damage or broken springs.

4 It is unlikely that the automatic advance unit will need to be dismantled, unless replacement parts have to be fitted.

10 Sparking plug: checking and resetting gap

1 A 10 mm NGK sparking plug is fitted to all C50 and C70 models as standard, the grade depending on the model designation. Refer to the Specifications Section heading this Chapter for the recommended grades. The C90 model has a 12 mm plug.

2 All models use a sparking plug with a 12.7 mm reach which should be gapped at 0.024 in. Always use the grade of plug recommended or the exact equivalent in another manufacturer's range.

3 Check the gap at the plug points during every six monthly or 3,000 mile service. To reset the gap, bend the outer electrode to bring it closer to the central electrode and check that a 0.024 inch feeler blade can be inserted. Never bend the central electrode, otherwise the insulator will crack, causing engine damage if particles fall in whilst the engine is running.

4 The condition of the sparking plug electrodes and insulator can be used as a reliable guide to engine operating conditions. See accompanying diagrams.

5 Always carry a spare sparking plug of the correct grade. In the rare event of plug failure it will enable the engine to be restarted.

6 Never over-tighten a sparking plug, otherwise there is risk of stripping the threads from the cylinder head, particularly those cast in light alloy. The plug should be sufficiently tight to seat firmly on the copper sealing washer. Use a spanner that is a good fit, otherwise the spanner may slip and break the insulator.

7 Make sure the plug insulating cap is a good fit and free from cracks. This cap contains the suppressor that eliminates radio and TV interference.

11 Fault diagnosis: ignition system

Symptom	Cause	Remedy
Engine will not start	No spark at plug	Try replacement plug if gap correct. Check whether contact breaker points are opening and closing, also whether they are clean. Check whether points arc when separated. If so, renew condenser. Check ignition switch and ignition coil. Battery discharged. Switch off all lights and use emergency start.
Engine starts but runs erratically	Intermittent or weak spark	Try replacement plug. Check whether points are arcing. If so, replace condenser. Check accuracy of ignition timing. Low output from flywheel magneto generator or imminent breakdown of ignition coil.
	Automatic advance unit stuck or damaged	Check unit for freedom of action and broken springs.

Chapter 5 Frame and forks

Contents

General description ... 1	Centre stand and prop stand: inspection ... 9
Front forks: removal from frame ... 2	Footrests: inspection and renovation ... 10
Front forks: dismantling ... 3	Speedometer: removal and replacement ... 11
Steering head bearings: examination and renovation ... 4	Speedometer cable: inspection and maintenance ... 12
Steering head lock ... 5	Dual seat: removal ... 13
Frame assembly: inspection and renovation ... 6	Cleaning the plastic moulding ... 14
Swinging arm rear fork: dismantling inspection and reassembly ... 7	Cleaning: general ... 15
Rear suspension units: dismantling and inspection ... 8	Fault diagnosis: frame and fork assembly ... 16

Specifications

Frame	Spine type, scooter style
Front suspension	Hydraulically damped leading link forks
Rear suspension	Pressed steel swinging arm controlled by hydraulically damped suspension units
Caster angle	63°
Trail	75 mm (2.95 inch)
Maximum lock	45° C50 and C70 models
	43° C90 model

1 General description

The frame used on these models is of the scooter style spine type ie. it has a single tube holding the steering head mounted low enough to step over. This is why the models are sometimes referred to as step-thru models. The forks are of the leading link variety, with hydraulic damping. The rear suspension is of the swinging arm type, controlled by two suspension units also with hydraulic damping.

The C70 model has, in addition to the normal centre stand, a prop stand mounted on the footrest bar.

2 Front forks: removal from frame

1 It is extremely unlikely that the front forks will need to be removed from the frame as a unit unless the steering head bearings give trouble or the forks are damaged in an accident.
2 Commence operations by placing the machine on the centre stand.
3 Remove the front wheel as described in Chapter 6.3.
4 Although not essential, it is advisable to remove the front mudguard for ease of handling. Remove the two self-locking nuts and their washers and the two upper fixing bolts and

2.4a Remove the self-locking nuts ...

2.4b ... and the bolts to release the mudguard

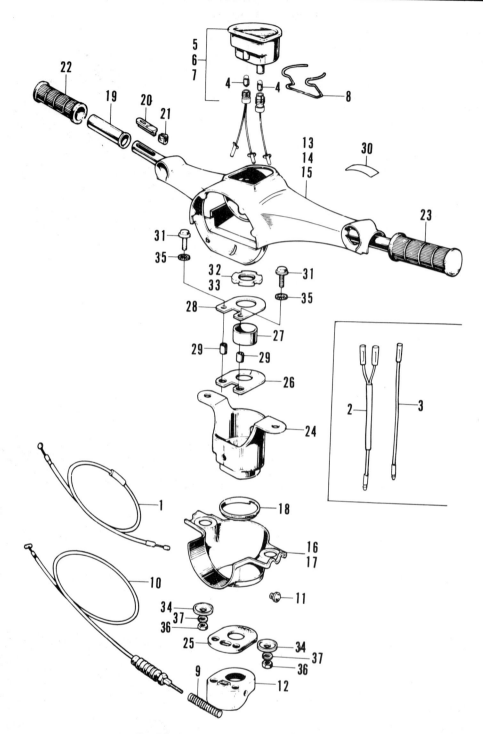

Fig. 5.1. Handlebar assembly - C70 model

1 Throttle cable	11 Plug	20 Sliding block	29 Spacer (2 off)
2 Earth wire	12 Fork top plate	21 Cable stop	30 Label
3 Earth wire	13 Handlebar assembly	22 Rubber grip	31 Bolt (2 off)
4 Bulb (2 off)	14 Handlebar assembly	23 Rubber grip	32 Special nut
5 Speedometer assembly	15 Handlebar assembly	24 Fork top bridge	33 Special nut
6 Speedometer assembly	16 Bottom cover	25 Top bridge rubber	34 Cup washer (2 off)
7 Speedometer assembly	17 Bottom cover	26 Top bridge rubber	35 Washer (2 off)
8 Spring	18 Rubber ring	27 Spacer	36 Nut (2 off)
9 Return spring	19 Twistgrip sleeve	28 Top bridge plate	37 Washer (2 off)
10 Front brake cable			

Chapter 5: Frame and forks

washers. Pull the mudguard off the pivot bolts and clear of the machine, ensuring that the front brake and speedometer cables slide out of the moulded slot.

5 Remove the two bolts and the headlamp unit. Disconnect the headlamp wires and put the headlamp in a safe place.

6 Unscrew the top of the carburettor and disconnect the cable by unhooking the throttle slide and pulling off the carburettor components.

7 Remove the two nuts and their washers from the underside of the handlebar fairing. The handlebars can now be eased up to allow the remainder of the wires to be disconnected at their snap connectors. All the wires are colour coded, to make re-connection easy. The handlebars, complete with the fairing, speedometer and cables, can now be pulled clear.

8 Remove the two bolts and lockwashers and the slotted nut to release the top bridge plate. Remove the plate, the spacers, and the top mounting rubber. The top bridge will now slide off. The handlebar lower cover with its sealing ring can be lifted clear. The lower mounting rubber can be removed and the two 8 mm bolts, which will permit the fork top plate to be removed.

9 Support the fork legs, to prevent them dropping prematurely, and remove the steering head column nut.

10 Remove the top cone to reveal 21 ball bearings in the top race, which can then be lifted out with a magnet or a greased screwdriver.

11 As the forks are now lowered, the 21 ball bearings in the lower race will be displaced and once these have been collected, the fork assembly can be pulled clear of the frame.

12 The bottom cone, the dust seal and the dust seal holder, can be removed from the fork assembly, if required.

2.7a Remove the two nuts and washers ...

2.7b ... and lift handlebars partly off

3 Front forks: dismantling

1 If only the fork leading links and suspension units are to be removed, without disturbing the head races, remove the front wheel as described in Chapter 6.3 and the self-locking nut and washer that retains the bottom of the mudguard.

2 Continue from here for further dismantling, irrespective of whether the forks are on or off the machine. Remove the bottom pivot bolt that was fitted with the self-locking nut. Remove the spring clip, the nut, the cup washer and the rubber washer, from the top of the suspension unit, and pull the link and suspension unit out of the forks.

3 The rubber rebound stop need not be removed unless it is damaged or perished, in which case it can be withdrawn by removing the nut and bolt that passes through it.

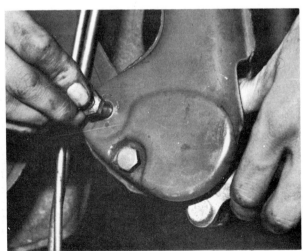

3.2a Remove rearmost bolt to release leading link ...

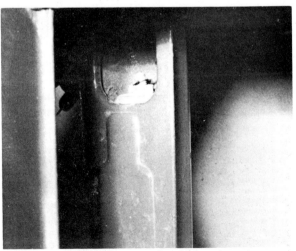

3.2b ... and the spring clip, nut and washer to release the suspension unit

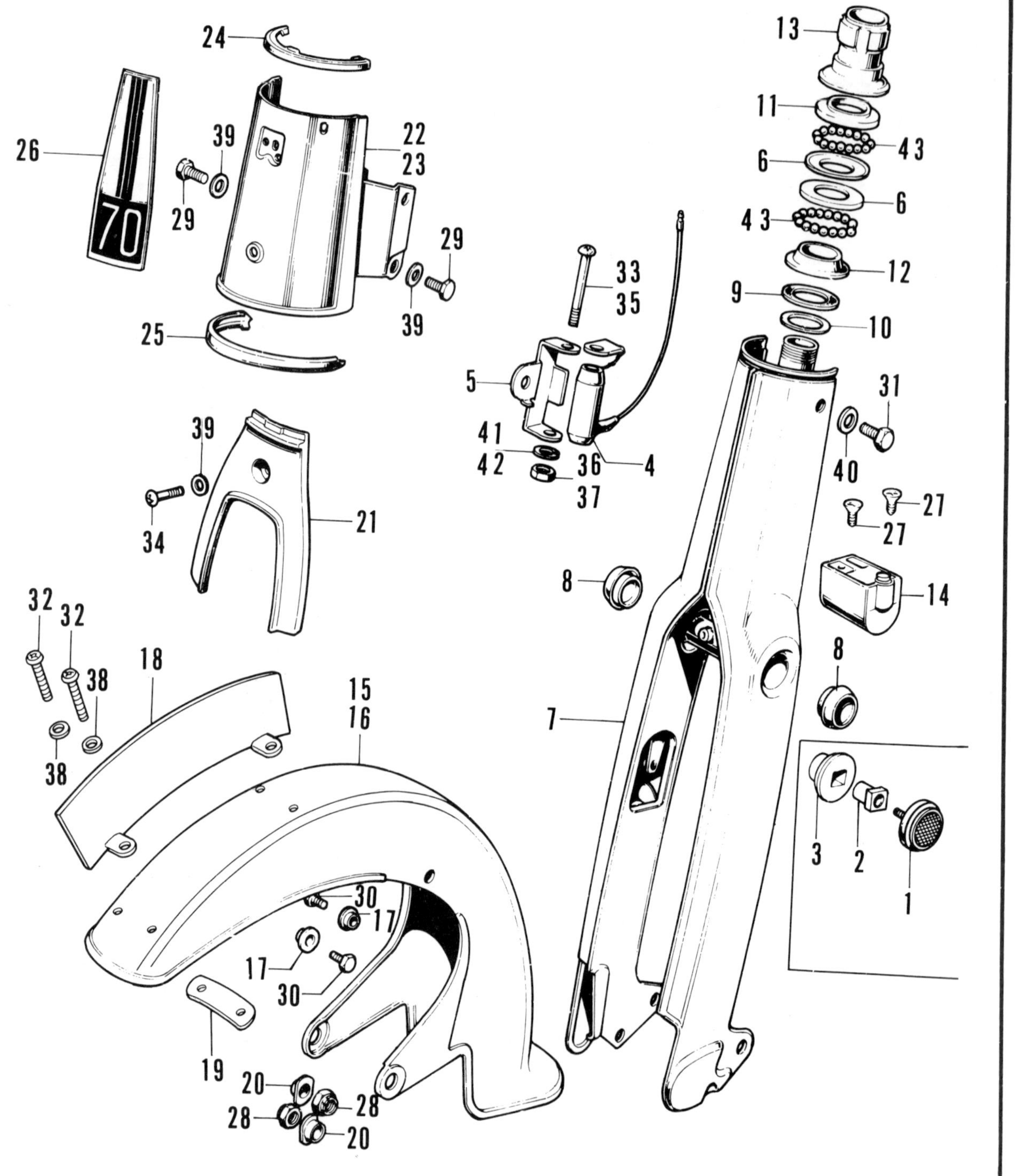

Fig. 5.2. Steering head assembly - C70 model

1 Reflector (2 off)	12 Bottom cone	23 Top cover	34 Screw
2 Setting piece (2 off)	13 Adjusting nut	24 Beading	35 Screw
3 Reflector base (2 off)	14 Steering lock	25 Beading	36 Nut
4 Resistor assembly	15 Front mudguard	26 Emblem	37 Nut
5 Mounting plate	16 Front mudguard	27 Screw (2 off)	38 Washer (4 off)
6 Ball race cup (2 off)	17 Spacer (2 off)	28 Self-locking nut (2 off)	39 Washer (3 off)
7 Front fork	18 Front number plate*	29 Bolt (2 off)	40 Washer (2 off)
8 Blanking cap (2 off)	19 Nut plate (2 off)	30 Bolt (2 off)	41 Washer
9 Dust seal	20 Spacer (2 off)	31 Bolt (2 off)	42 Washer
10 Dust seal washer	21 Cover	32 Screw (4 off)	43 Ball bearing (42 off)
11 Top cone	22 Top cover	33 Screw	

* not required on UK models

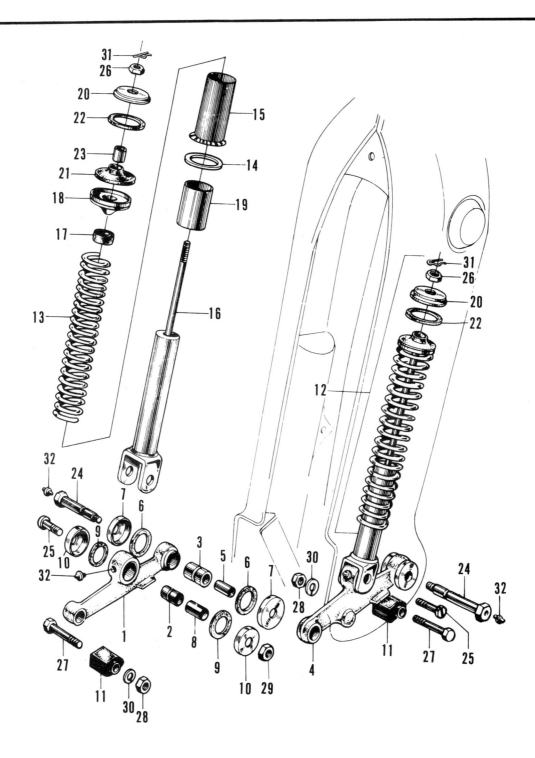

Fig. 5.3. Front suspension - C70 model

1. Right-hand leading link
2. Bush (2 off)
3. Bush (2 off)
4. Left-hand leading link
5. Spacer (2 off)
6. Dust seal (4 off)
7. Dust seal cap (4 off)
8. Spacer (2 off)
9. Dust seal (4 off)
10. Dust seal cap (4 off)
11. Rebound stop (2 off)
12. Front shock absorber assembly (2 off)
13. Spring (2 off)
14. Spring register (2 off)
15. Spring guide (2 off)
16. Damper assembly (2 off)
17. Rubber stop (2 off)
18. Tapped spring register (2 off)
19. Spacer (2 off)
20. Cup washer (2 off)
21. Rubber mount (2 off)
22. Rubber ring (2 off)
23. Spacer (2 off)
24. Bolt (2 off)
25. Bolt (2 off)
26. Nut (2 off)
27. Bolt (2 off)
28. Nut (2 off)
29. Nut (2 off)
30. Washer (2 off)
31. Spring clip (2 off)
32. Grease nipple (4 off)

4 To detach the bottom link from the suspension unit, remove the pivot bolt and nut.
5 If play is evident between any of the pivot pins and their respective bushes, both components will need to be renewed.
6 The dust seals and covers will need to be removed to gain access to the bushes and where the arm is peened over to retain the cover, care should be taken not to destroy the latter.
7 The bush is a press fit in the pivot eye, and the new bush can be used to press out the old one.
8 The suspension units can be dismantled partially if it is desired to remove the spring. Pull off the rubber disc and the spacer, unscrew the threaded spring register and remove the spring, the rubber stop, the spring guide, the thrust washer and the spacing collar. Renew the spring if it is below 120 mm (4.72 in) in length.
9 Note that the damper units are sealed and cannot be dismantled further. If the units leak or if the damping action is lost, replacement of the unit is the only satisfactory answer.

4 Steering head bearings: examination and renovation

1 Before commencing to reassemble the forks, inspect the steering head races. The ball bearing tracks should be polished and free from indentations or cracks. If signs of wear or damage are evident, the cups and cones must be replaced. They are a tight push fit and need to be drifted out of position.
2 Ball bearings are cheap. Each race has 21 x 6 mm ball bearings, which should be renewed if the originals are marked or discoloured. To hold the steel balls in position during the re-attachment of the forks, pack the bearings with grease.
3 The forks are reassembled and attached to the frame by following a reversal of the dismantling procedure. Do not over-tighten the steering head bearings, otherwise the handling characteristics of the machine will be affected. It is possible to over-tighten and create a load of several tons on the steering head bearings, even though the handlebars appear to turn with relative ease. As a guide, only very slight pressure should be needed to start the front wheel turning to either side under its own weight, when it is raised clear of the ground. Check also that the bearings are not too slack; there should be no discernable movement of the forks, in the fore and aft direction.

3.4 Remove the pivot bolt to release the link

3.6 Dust cap and felt washer seal the bearing

5.1 Steering lock is mounted under steering head

7.5 Rubber bushes can be driven out, if perished

Chapter 5: Frame and forks

5 Steering head lock

A steering head lock is attached to the lower fork yoke by two crosshead screws. The tongue of the lock engages with a hole drilled in a plate attached to the steering head column, so that the machine can be left unattended with the handlebars on full left lock. If the lock malfunctions, it must be replaced.

6 Frame assembly: inspection and renovation

1 The frame assembly is unlikely to require attention unless it has been damaged in an accident. Replacement of the complete unit is the only satisfactory course of action when the frame is out of alignment, if only because the special jigs and mandrels essential for resetting the frame will not be available.
2 After a machine has covered an extensive mileage, it is advisable to inspect the frame for signs of cracking or splitting in the vicinity of the engine mounting points. Repairs can be effected by welding.
3 Frame alignment should be checked when the machine is complete. The accompanying diagram shows how a board placed each side of the rear wheel can be used as a guide to alignment. It is, of course, necessary to ensure that both wheels are centrally disposed within their respective forks before carrying out this check.

7 Swinging arm rear fork: dismantling, inspection and reassembly

1 The rear fork of the frame assembly pivots on the rear fork pivot bolt which is carried in rubber bushes. Its action is controlled by two hydraulically-damped rear suspension units that each carry a compression spring, connected between the swinging arm fork and the frame, one on each side of the machine.
2 To detach the rear fork, remove the rear wheel, as described in Chapter 6.9. Remove the upper and lower halve of the rear chaincase and remove the final drive chain. Remove the final drive sprocket as described in Chapter 6.12.
3 Unscrew the 10 mm nut attaching each of the rear suspension units to the frame. Remove the domed nut that secures the lower eye of each unit to the swinging arm fork.
4 Unscrew the rear fork pivot nut and withdraw the fork pivot bolt. The swinging arm fork can now be removed from the frame.
5 Check the condition of the rubber bushes in the swinging arm fork pivot. A rubber bonded bush is fitted to each side, which can be pressed out if renewal is necessary. Replace the bushes if they show signs of damage or ageing, or if the pivot bolt is a slack fit. Check also the pivot bolt.
6 Place a straight metal rod across the open ends of the swinging arm fork, where the rear wheel spindle is normally located. Check for twist or deformation. If the amount of twist is more than 1 mm, the swinging arm fork should be replaced. A twisted rear fork will throw the wheels out of track and give the machine poor handling characteristics.
7 To reassemble the swinging arm fork assembly, reverse the procedure detailed above.

8 Rear suspension units: dismantling and inspection

1 The rear suspension units can be partially dismantled to gain access to the springs. Unscrew the upper eye of the unit from the damper rod after slackening the locknut. When the pivot eye has been removed, the top shroud can be pulled off and the spring and spring guide detached. If the free length of the spring is below 200 mm, it must be renewed.
2 If the suspension units show signs of leakage or if the damping action is no longer evident, the units should be renewed. Always renew both, as a matched pair, in the interests of good roadholding. The damper unit is sealed and cannot be repaired.

9 Centre stand and prop stand: inspection

1 The centre stand is attached to the lower extremities of the frame unit, to provide a convenient means of parking the machine on level ground. It pivots on the hollow tube that carries the rear brake pedal, which is retained in position by a split pin. A return spring retracts the stand when the machine is pushed forward, so that it can be wheeled prior to riding.
2 On the C70 model there is a prop stand fitted onto the footrest bar, also with a return spring.
3 The condition of the return springs and their return action should be checked regularly. If a stand falls whilst the machine is in motion, it may catch in some obstacle in the road and unseat the rider.

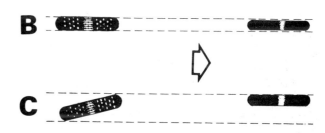

Fig. 5.4. Checking wheel alignment

A and C — Incorrect
B — Correct

9.2 Check the condition of the prop stand spring

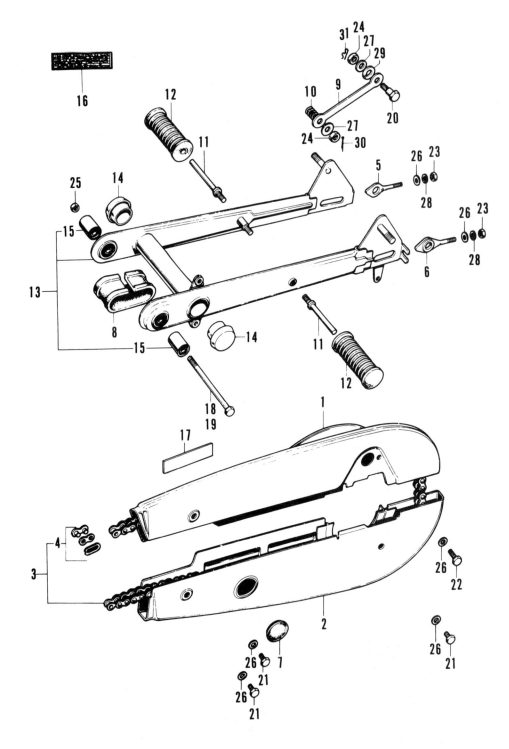

Fig. 5.5. Swinging arm and chaincase - C70 model

1 Upper half of chaincase	9 Anchor arm	17 Label	25 Self-locking nut
2 Bottom half of chaincase	10 Spring	18 Spindle bolt	26 Washer (6 off)
3 Chain	11 Pillion footrest (2 off)	19 Spindle bolt	27 Washer (2 off)
4 Spring link	12 Footrest rubber (2 off)	20 Shouldered bolt	28 Washer (2 off)
5 Chain adjuster	13 Swinging arm assembly	21 Bolt (3 off)	29 Washer
6 Chain adjuster	14 Blanking cap (2 off)	22 Bolt	30 Split pin
7 Blanking cap	15 Rubber bush (2 off)	23 Nut (2 off)	31 Spring clip
8 Protection strip	16 Label	24 Nut (2 off)	

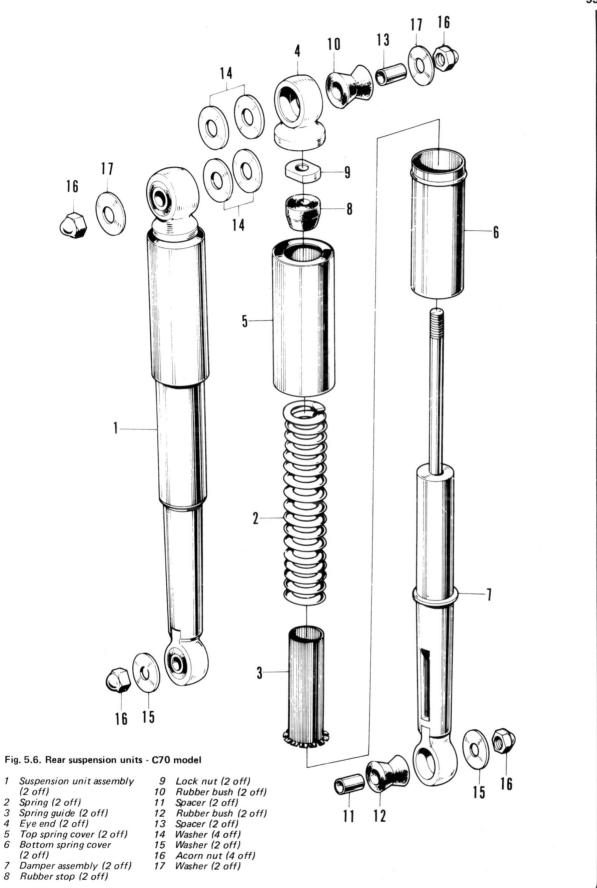

Fig. 5.6. Rear suspension units - C70 model

1 Suspension unit assembly (2 off)
2 Spring (2 off)
3 Spring guide (2 off)
4 Eye end (2 off)
5 Top spring cover (2 off)
6 Bottom spring cover (2 off)
7 Damper assembly (2 off)
8 Rubber stop (2 off)
9 Lock nut (2 off)
10 Rubber bush (2 off)
11 Spacer (2 off)
12 Rubber bush (2 off)
13 Spacer (2 off)
14 Washer (4 off)
15 Washer (2 off)
16 Acorn nut (4 off)
17 Washer (2 off)

10.1 Four bolts hold the footrest bar

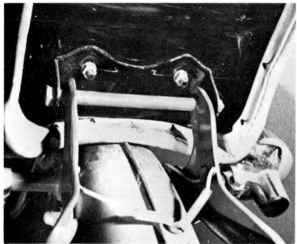

13.1 Two nuts hold the dual seat

10 Footrests: inspection and renovation

1 The footrest bar is attached to the bottom of the crankcase of the engine unit by four 8 mm bolts and washers. The bar is malleable and is likely to become bent if the machine is dropped.
2 To straighten the bar first remove it from the machine and detach the footrest rubbers. It can then be bent straight in a vice, using a blow lamp to warm the tube if the bend is severe. Never attempt to straighten the bar whilst it is still attached to the crankcase, otherwise serious damage to the crankcase casting may result.

11 Speedometer: removal and replacement

1 A speedometer of the magnetic type is fitted to these Honda models. It contains also the odometer for recording the total mileage covered by the machine, a small bulb for illuminating the dial during the hours of darkness, and the neutral indicator lamp.
2 The speedometer is held in position by a tension spring. To remove the speedometer, remove the two bolts and the headlamp unit, and remove the two nuts and washers from the underside of the handlebar fairing. Ease the cables into the fairing and lift the fairing sufficiently to unscrew the speedometer cable. Ease the fairing higher and release the clip that holds the speedometer head. The speedometer can now be lifted clear, from the top.
3 Although a speedometer on a machine of less than 100 cc capacity is not a statutory requirement in the UK, if one is fitted it must be in good working order. Reference to the mileage reading shown on the odometer is a good way of keeping in pace with the routine maintenance schedules.
4 Apart from defects in the speedometer drive or in the drive cable itself, a speedometer that malfunctions is difficult to repair. Fit a replacement or alternatively entrust the repair to an instrument repair specialist.

12 Speedometer cable: inspection and maintenance

1 It is advisable to detach the speedometer drive cable from time to time, in order to check whether it is adequately lubricated and whether the outer covering is compressed or damaged at any point along its run. A jerky or sluggish speedometer movement can often be attributed to a cable fault.
2 To grease the cable, withdraw the inner cable. After removing the old grease, clean with a petrol soaked rag and examine the cable for broken strands or other damage.
3 Re-grease the cable with high melting point grease, taking care not to grease the last six inches at the point where the cable enters the speedometer head. If this precaution is not observed, grease will work into the speedometer head and immobilise the movement.
4 If the speedometer and the odometer stop working, it is probable that the speedometer cable has broken. Inspection will show whether the inner cable has broken; if so, the inner cable alone can be renewed and reinserted in the outer covering after greasing. Never fit a new inner cable alone if the outer covering is damaged or compressed at any point along its run.

13 Dualseat: removal

1 The dualseat is removed by unscrewing the two bolts in the base of the seat, at the rear.
2 When the seat has been detached, the long bolt through the rear of the carrier can be removed.

14 Cleaning the plastic mouldings

1 The moulded plastic cycle parts, such as the front mudguard, and legshield cover will not respond to cleaning in the same way as the other metal parts as they are moulded in rigid polyethylene. It is best to wash these parts with a household detergent solution, which will remove oil and grease in a most effective manner.
2 Avoid the use of scouring powder as much as possible because this will score the surface of the mouldings and make them more receptive to dirt.

15 Cleaning: general

1 After removing all surface dirt with a rag or sponge that is washed frequently in clean water, the application of car polish or wax will give a good finish to the cycle parts of the machine, after they have dried thoroughly. The plated parts should require only a wipe over with a damp rag.
2 If possible, the machine should be wiped over immediately after it has been used in the wet, so that it is not garaged in damp conditions that will promote rusting. Make sure to wipe the chain and if necessary re-oil it, to prevent water from entering the rollers and causing harshness with an accompanying rapid rate of wear. Remember there is little chance of water entering the control cables if they are lubricated regularly, as recommended in the Routine Maintenance Section.

Chapter 5: Frame and forks

16 Fault diagnosis: Frame and fork assembly

Symptom	Cause	Remedy
Machine is unduly sensitive to road surface irregularities	Fork and/or rear suspension units damping ineffective ive	Renew suspension unit.
Machine rolls at low speeds	Steering head bearings overtight or damaged	Slacken bearing adjustment. If no improvement, dismantle and inspect head bearings.
Machine tends to wander; steering is imprecise	Worn swinging arm suspension bearings	Check and if necessary renew pivot bolt and bushes.
Fork action stiff	Fork legs have twisted	Check alignment.
Forks judder when front brake is applied	Worn bushes in fork assembly Steering head bearings too slack	Strip forks and renew bushes. Readjust to take up play.
Wheels seem out of alignment	Frame distorted as result of accident damage	Check frame after stripping out. If bent, replacement frame is necessary.

Chapter 6 Wheels, brakes and tyres

Contents

General description ... 1	reassembly ... 10
Front wheel: examination and renovation ... 2	Rear wheel bearings: examination and replacement ... 11
Front wheel: removal ... 3	Rear wheel sprocket: removal, examination and replacement ... 12
Front brake assembly: examination, renovation and reassembly ... 4	Rear wheel shock absorber assembly: examination and replacement ... 13
Front wheel bearings: examination and replacement ... 5	Rear wheel: reassembly ... 14
Speedometer drive gears: examination and replacement ... 6	Front and rear brakes: adjustment ... 15
Front wheel: replacement ... 7	Final drive chain: examination and lubrication ... 16
Rear wheel: examination and renovation ... 8	Tyres: removal and replacement ... 17
Rear wheel: removal ... 9	Fault diagnosis: wheels, brakes and final drive ... 18
Rear brake assembly: examination, renovation and	

Specifications

Wheels ...	17 inch diameter front and rear
Tyres ...	2.25 inch x 17 inch front and rear (C50 and C70 models)
	2.50 inch x 17 inch front and rear (C90 model)
Tyre pressures	
Front ...	26 psi (1.8 kg/cm^2)
Rear solo ...	29 psi (2.0 kg/cm^2)
with pillion ...	32 psi (2.2 kg/cm^2)
Brakes ...	Internally expanding 110 mm diameter front and rear
Chain size ...	½ inch x ¼ inch all models
Chain length ...	98 links (C50 and C70 models)
	100 links (C90 model)
Engine sprocket ...	13 teeth (C50 model)
	14 teeth (C70 and C90 models)
Rear wheel sprocket ...	41 teeth (C50 model)
	39 teeth (C70 model)
	40 teeth (C90 model)

1 General description

Both wheels are of 17 inch diameter and carry tyres of 2.25 inch section (2.50 inch section, 90 cc models), a ribbed tyre at the front and a block tread at the rear. Steel wheel rims are used in conjunction with cast aluminium alloy hubs, each hub containing a 110 mm internal expanding brake. The wheels are not interchangeable because the rear wheel incorporates a special cush drive arrangement to act as a transmission shock absorber. Both wheels are quickly detachable; the rear wheel can be removed from the frame without disturbing the rear wheel sprocket or the final drive chain.

2 Front wheel: examination and renovation

1 Place the machine on the centre stand so that the front wheel is raised clear of the ground. Spin the wheel and check for rim alignment or run-out. Small irregularities can be corrected by tightening the spokes in the area affected, although a certain amount of experience is advisable if over-correction is to be avoided.

2 Any flats in the wheel rim should be evident at the same time. These are much more difficult to remove and in most cases the wheel will need to be rebuilt on a new rim. Apart from the effect on stability, there is greater risk of damage to the tyre

Chapter 6: Wheels, brakes and tyres

bead and walls if the machine is run with a deformed wheel. In an extreme case the tyre can even separate from the rim.

3 Check for loose or broken spokes. Tapping the spokes is the best guide to tension. A loose spoke will produce a quite different sound and should be tightened by turning the nipple in an anticlockwise direction. Always recheck for run-out by spinning the wheel again.

4 If it is necessary to turn a spoke nipple an excessive amount to restore tension, it is advisable to remove the tyre and tube so that the end of the spoke that now protrudes into the wheel rim can be ground flush. If this precaution is not taken, there is danger of the spoke end chafing the inner tube and causing an eventual puncture.

3 Front wheel: removal

1 Commence operations by placing the machine on the centre stand, on level ground.
2 Disconnect the speedometer cable by unscrewing it from the brake backplate and disconnect the front brake cable.
3 On the early C90 models there is a separate torque arm, which needs to be disconnected.
4 Withdraw the split pin from the front wheel spindle nut and remove the nut and washer. The front wheel spindle can now be pulled out, releasing the wheel complete with brake assembly. Ensure that the machine is well supported so that it does not topple forward once the wheel is removed.

4 Front brake assembly: examination, renovation and reassembly

1 To remove the brake assembly, the backplate lifts straight out of the brake drum.
2 Examine the brake linings. If they are wearing thin or unevenly, the brake shoes should be renewed. The linings are bonded on and cannot be replaced as a separate item.
3 To remove the brake shoes from the brake plate assembly, arrange the operating lever so that the brakes are in the 'full on' position and then pull the shoes apart whilst lifting them upward in the form of a 'V'. When they are clear of the brake plate, the return springs can be removed and the shoes separated.
4 Before replacing the brake shoes, check that the brake operating cam is working smoothly and is not binding in its pivot. The cam can be removed by withdrawing the retaining nut on the operating arm and pulling the arm off the shaft. Before

2.1 Badly buckled wheel rim

3.2 Disconnect the speedometer and brake cables

3.3 Early models have an anchor arm

4.1 The brake assembly lifts out of the drum

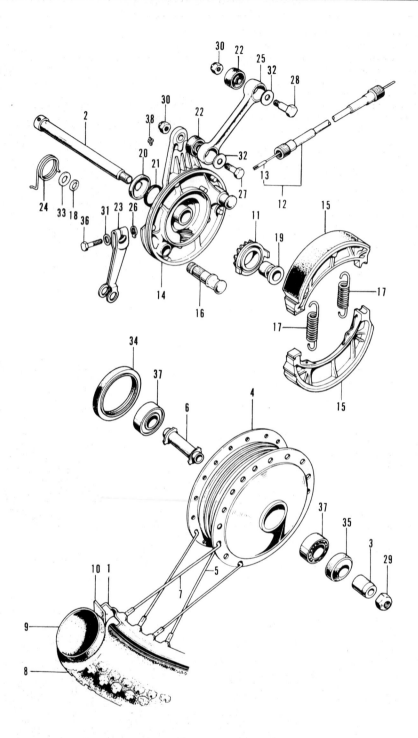

Fig. 6.1. Front wheel assembly - C90 model

1	Wheel rim	11	Speedometer gear	21	Dust seal	31	Washer
2	Spindle	12	Speedometer cable	22	Rubber bush (2 off)	32	Washer (2 off)
3	Spacer	13	Speedometer cable	23	Brake lever arm	33	Washer
4	Wheel hub	14	Brake backplate	24	Brake arm return spring	34	Oil seal
5	Spoke (18 off)	15	Brake shoe (2 off)	25	Anchor arm	35	Oil seal
6	Bearing spacer	16	Operating cam	26	Special nut	36	Bolt
7	Spoke (18 off)	17	Brake shoe spring (2 off)	27	Bolt	37	Ball bearing (2 off)
8	Tyre	18	Dust seal	28	Bolt	38	Grease nipple
9	Inner tube	19	Spacer	29	Spindle nut		
10	Rim tape	20	Dust seal cap	30	Self-locking nut (2 off)		

Chapter 6: Wheels, brakes and tyres

removing the arm, it is advisable to mark its position in relation to the shaft, so that it can be relocated correctly. The shaft should be greased prior to reassembly and also a light smear of grease placed on the faces of the operating cam.

5 Check the inner surface of the brake drum on which the brake shoes bear. The surface should be smooth and free from score marks or indentations, otherwise reduced braking efficiency will be inevitable. Remove all traces of brake lining dust and wipe with a clean rag soaked in petrol to remove any traces of grease or oil.

6 If the brake drum has become scored, special attention is required. It is possible to skim a brake drum in a lathe provided the score marks are not too deep. Under these circumstances, packing will have to be added to the ends of the brake shoes, to compensate for the amount of metal removed from the surface of the drum.

7 To reassemble the brake shoes on the brake plate, fit the return springs first and then force the shoes apart, holding them in a 'V' formation. If they are now located with the brake operating cam and pivot they can usually be snapped into position by pressing downward. Never use excessive force, otherwise there is risk of distorting the shoes permanently.

5 Front wheel bearings: examination and replacement

1 The front wheel bearings are of the ball journal type and are not adjustable. If the bearings are worn, indicated by side play at the wheel rim, the bearings must be renewed.

2 Access to the wheel bearings is gained when the brake plate has been removed from the front wheel. There is an oil seal in front of the bearing on the brake drum side, to prevent grease from reaching the brake operating parts. This seal should be prised out of position and a new replacement obtained.

3 The wheel bearings are a drive fit in the hub. Use a double diameter drift to displace the bearings from the hub, working from each side of the hub. When the first bearing emerges from the hub the hollow distance collar that separates them can be removed.

4 Remove all the old grease from the hub and bearings, giving the latter a final wash in petrol. Check the bearings for play or signs of roughness when they are turned. If there is any doubt about their condition, renew them.

5 Before replacing the bearings, first pack the hub with new grease. Then grease both bearings and drive them back into position with the same double diameter drift, not forgetting the

4.4 Punch marks on brake lever and operating spindle allow correct reassembly

4.5 Clean the brake drum to remove any dust

4.7 Snap brake shoes back into position, as shown

5.3 Use a suitable drift for removing and replacing bearings

5.5a Refit the new bearings ...

5.5b ... and use care when fitting the oil seal

6.4 Small screw retains worm gear

6.5 Speedometer drive gear tongues engage with slots in hub

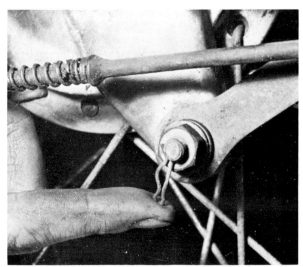
9.1a Remove the spring clip and nut ...

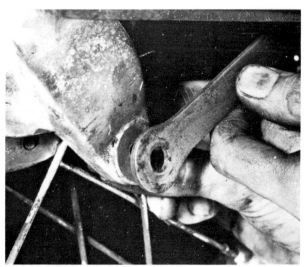
9.1b ... and pull the anchor arm free

Chapter 6: Wheels, brakes and tyres

distance collar that separates them. Fit the replacement oil seals in front of the bearing on the brake drum side.

6 Speedometer drive gears: examination and replacement

1 The speedometer drive gears are housed in the front brake backplate and are driven by two tongues that engage with slots in the hub.
2 The drive gears should be checked for wear or broken teeth and renewed if necessary.
3 The large gear with its two drive tongues simply lifts out of the backplate.
4 To renew the small worm gear it is necessary to remove the very small screw before prising it out, together with its bushes.
5 Reassembly of the gears is by reversing the above procedure, ensuring that the gear tongues engage with the hub slots. Thoroughly grease the gears and ensure that the oil seal is in good condition.

7 Front wheel: replacement

1 To replace the front wheel, reverse the removal procedure and ensure that the peg on the forks locates in the slot in the brakeplate, or the anchor arm is fixed securely. This cannot be overstressed as failure to anchor the brakeplate will cause the brake to lock on when it is first applied.
2 Reconnect the front brake and check that the brake functions correctly, especially if the adjustment has been altered or the brake operating arm has been removed and replaced during the dismantling operation.
3 Reconnect the speedometer drive.
4 On early C90 models do not omit to replace and tighten the torque arm.

8 Rear wheel: examination and renovation

1 Place the machine on the centre stand, so that the rear wheel is clear of the ground. Check the wheel for rim alignment, damage to the rim or loose or broken spokes, by following the procedure adopted for the front wheel in Section 2 of this Chapter.

9 Rear wheel: removal

1 To remove the rear wheel, place the machine on the centre stand so that the wheel is raised clear of the ground. Remove the brake adjusting nut and separate the brake rod from the operating arm of the rear brake. Remove the clip, slacken and remove the nut and bolt from the rear brake torque arm, at the brake plate anchorage.
2 Remove the split pin, the centre of the two rear wheel spindle nuts and pull the spindle from the hub. Remove the distance pieces if they have not fallen clear and disengage the hub from the cush drive assembly by pulling the wheel towards the right-hand side of the machine. The wheel can now be removed from the frame complete with the rear brake assembly, leaving the rear sprocket and final drive chain in position.

10 Rear brake assembly: examination, renovation and reassembly

As the rear brake is identical to the front brake, the advice given in Section 4 of this Chapter will apply.

11 Rear wheel bearings: examination and replacement

As the bearing layout for the hubs is identical, the procedure

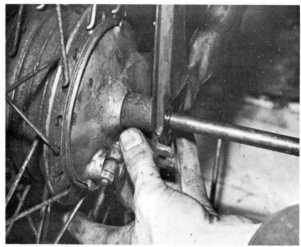
9.2 Pull out the spindle and remove the spacer

12.3 Remove the nut to release the sprocket assembly

described in Section 5 of this Chapter will apply for each model.

12 Rear wheel sprocket: removal, examination and replacement

1 It is unlikely that the sprocket will require renewal until a very substantial mileage has been covered. The usual signs of wear occur when the teeth assume a hooked or very shallow formation that will cause rapid wear of the chain. A worn sprocket must be renewed, together with the gearbox final drive sprocket and the chain. Always renew the final drive assembly as a complete set, otherwise rapid wear will occur as the result of running old and new parts together.
2 As the sprocket is left attached to the machine when the rear wheel is removed, the chainguard halves must be removed for access. Remove the two bolts on each half and pull the top half clear first.
3 Disconnect the chain at the spring link and pull it clear of the rear sprocket. Remove the stub axle nut and washer and pull the sprocket assembly clear of the machine. Prise down the two double tab washers, remove the four fixing bolts and lift the sprocket off.
4 To refit the sprocket, reverse the dismantling procedure.

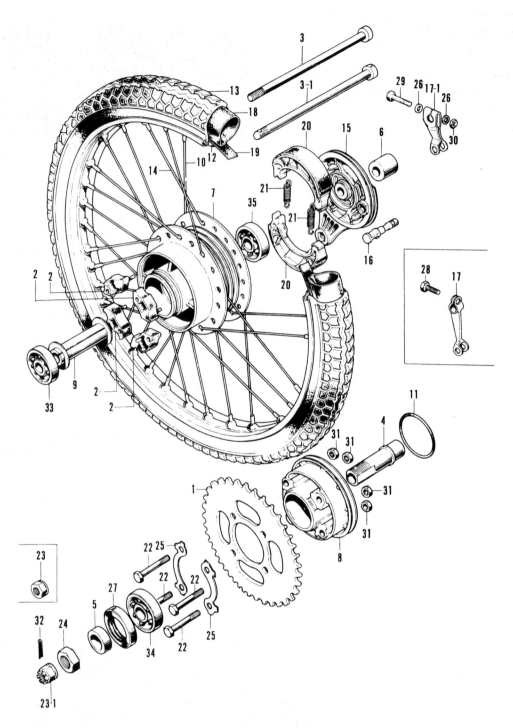

Fig. 6.2. Rear wheel assembly - C90 model

1 Sprocket	9 Bearing spacer	18 Inner tube	27 Oil seal
2 Shock absorber rubber (4 off)	10 Spoke (18 off)	19 Rim tape	28 Bolt
	11 O-ring	20 Brake shoe (2 off)	29 Bolt
3 Wheel spindle	12 Wheel rim	21 Brake shoe spring (2 off)	30 Nut
4 Stub axle	13 Tyre	22 Bolt (4 off)	31 Nut (4 off)
5 Spacer	14 Spoke (18 off)	23 Spindle nut	32 Split pin
6 Spacer	15 Brake backplate	24 Stub axle nut	33 Ball bearing
7 Wheel hub	16 Operating cam	25 Tab washer (2 off)	34 Ball bearing
8 Shock absorber hub	17 Brake lever arm	26 Washer (2 off)	35 Ball bearing

13 Rear wheel shock absorber assembly: examination and replacement

1 The shock absorber assembly is removed from the machine as described in the preceding Section.
2 The shock absorber assembly has a ball journal bearing and if worn the following procedure should be followed.
3 Tap out the stub axle, remove the spacer from the centre of the oil seal, prise out the oil seal and drive the bearing out of the housing.
4 To replace the bearing, reverse the above procedure, ensuring that the bearing is well greased and the oil seal is in good condition.
5 The shock absorber rubbers should be checked for any damage or deterioration. All oil or grease should be wiped away as this may cause premature deterioration.
6 To refit the shock absorber assembly reverse the dismantling procedure described in Sections 9 and 12.

14 Rear wheel: reassembly

1 To refit the rear wheel, reverse the removal procedure, ensuring that the brake anchor arm is securely fitted. If the anchor arm becomes detached, the rear brake will lock in the pull-on position immediately it is applied and may give rise to a serious accident.
2 Before fully tightening all the nuts, ensure that the final chain tension and the brake adjustment is correct.
3 Check also whether the wheel alignment is correct as described in Chapter 5, Section 6, paragraph 3.
4 Do not omit to replace the split pin that passes through the rear wheel spindle nut.

15 Front and rear brakes: adjustment

1 The front brake adjuster is located on the front brake cable. The brake should be adjusted so that the wheel is free to revolve before pressure is applied to the handlebar lever, and is applied fully before the handlebar lever touches the handlebar.
2 The rear brake is adjusted by means of the adjusting nut on the end of the brake cable or rod. Adjustment is largely a matter of personal choice, but excessive travel of the footbrake pedal should not be necessary before the brake is applied fully.

13.3 Remove the stub axle

13.4a Refit the new bearing ...

13.4b ... and refit the oil seal

13.5 Check the condition of the cush drive rubbers

Chapter 6: Wheels, brakes and tyres

16.2 Check for correct chain tension in middle of bottom run

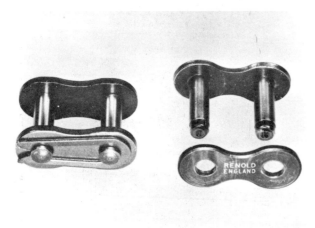

16.5 There is a chain of British manufacture available when renewal is necessary

16.7 Ensure that the spring clip is fitted the correct way round

3 Efficient brakes depend on good leverage of the operating arm. The angle between the brake operating arm and the cable or rod should never exceed 90° when the brake is fully applied.
4 Check that the brakes pull off correctly when the lever and pedal are released. Sluggish action can usually be traced to a broken return spring on the brake shoes or a tendency for the operating cam to bind in its bush.

16 Final drive chain: examination and lubrication

1 The final drive chain is fully enclosed within a chaincase. Periodically, the tension of the chain will need to be adjusted, to compensate for wear. This is accomplished by sliding the rear wheel backwards in the ends of the rear fork, using the drawbolt adjusters provided. The rear wheel spindle nuts must be slackened before the drawbolts can be turned; also the torque arm bolt on the rear brake plate.
2 The chain is in correct adjustment if there is from 1 - 2 cm free play in the middle of the lower run. An inspection plug in the lower section of the full chaincase permits access to the chain to check whether the tension is correct.
3 Always adjust the chain adjusters an identical amount, otherwise the rear wheel will be thrown out of alignment. If in doubt about the correctness of wheel alignment, use the technique described in Chapter 5, Section 6, paragraph 3.
4 After a period of running, the chain will require lubrication. Lack of oil will accelerate the rate of wear of both chain and sprockets, leading to harsh transmission. The application of engine oil from an oil can will serve as a satisfactory lubricant, but it is preferable to remove the chain at regular intervals and immerse it in a molten lubricant such as Linklyfe, after it has been cleaned in a paraffin bath. This latter type of lubricant achieves better penetration of the chain links and rollers and is less likely to be thrown off when the chain is in motion. An equally effective and less messy alternative is a spray-on lubricant of the aerosol type, such as Castrol Chain Lubricant.
5 The chain fitted as standard is of Japanese manufacture. When renewal is necessary, it should be noted that a Renold equivalent of British manufacture is available as an alternative. The size and length of chain is stated in the Specifications Section.
6 To check whether the chain requires replacement, lay it lengthwise in a straight line and compress it so that all play is taken up. Anchor one end and then pull on the other end to take up the end play in the other direction. If the chain extends by more than the distance between two adjacent rollers, it should be renewed in conjunction with the two sprockets. Note that this check should be made after the chain has been washed but before any lubricant is applied, otherwise the lubricant will take up some of the play.
7 When replacing the chain, make sure the spring link is correctly seated, with the closed end facing the direction of travel. Also make sure the chaincase is firmly secured.

17 Tyres: removal and replacement

1 At some time or other the need will arise to remove and replace the tyres, either as the result of a puncture or because a replacement is required to offset wear. To the inexperienced, tyre changing represents a formidable task, yet if a few simple rules are observed and the technique learned, the whole operation is surprisingly simple.
2 To remove the tyre from either wheel, first detach the wheel from the machine by following the procedure in Sections 3 or 9, depending on whether the front or the rear wheel is involved. Deflate the tyre by removing the valve insert and when it is fully deflated, push the bead of the tyre away from the wheel rim on both sides so that the bead enters the centre well of the rim. Remove the locking cap and push the tyre valve into the tyre itself.
3 Insert a tyre lever close to the valve and lever the edge of the

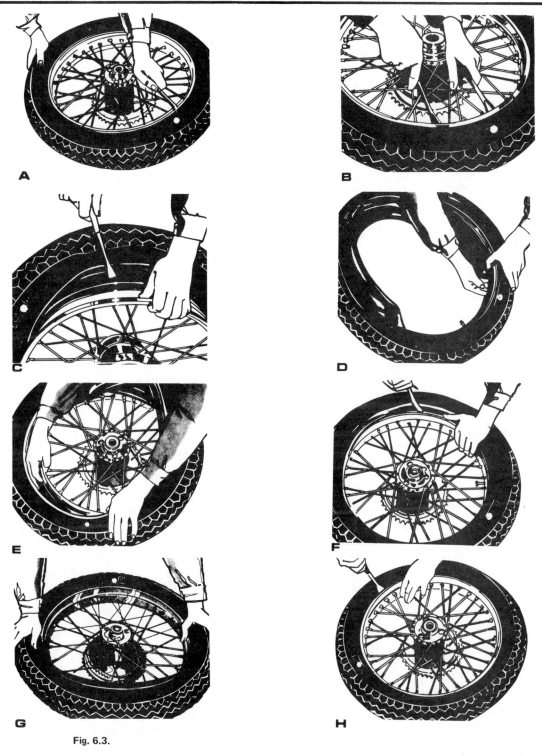

Fig. 6.3.

Tyre removal

- A Deflate inner tube and insert lever in close proximity to tyre valve
- B Use two levers to work bead over the edge of rim
- C When first bead is clear of rim, remove tyre as shown

Tyre fitting

- D Inflate inner tube and insert in tyre
- E Lay tyre on rim and feed valve through hole in rim
- F Work first bead over rim, using lever for final section
- G Use similar technique for second bead. Finish at tyre valve position
- H Push valve and tube up into tyre when fitting final section, to avoid trapping

tyre over the outside of the wheel rim. Very little force should be necessary; if resistance is encountered it is probably due to the fact that the tyre beads have not entered the well of the wheel rim all the way round the tyre.

4   Once the tyre has been edged over the wheel rim, it is easy to work around the wheel rim so that the tyre is completely free on one side. At this stage, the inner tube can be removed.

5   Working from the other side of the wheel, ease the other edge of the tyre over the outside of the wheel rim that is furthest away. Continue to work around the rim until the tyre is free completely from the rim.

6   If a puncture has necessitated the removal of the tyre, re-inflate the inner tube and immerse it in a bowl of water to trace the source of the leak. Mark its position and deflate the tube. Dry the tube and clean the area around the puncture with a petrol-soaked rag. When the surface has dried, apply the rubber solution and allow this to dry before removing the backing from the patch and applying the patch to the surface.

7   It is best to use a patch of the self-vulcanising type, which will form a very permanent repair. Note that it may be necessary to remove a protective covering from the top surface of the patch after it has sealed in position. Inner tubes made from synthetic rubber may require a special type of patch and adhesive, if a satisfactory bond is to be achieved.

8   Before replacing the tyre, check the inside to make sure the agent that caused the puncture is not trapped. Check also the outside of the tyre, particularly the tread area, to make sure nothing is trapped that may cause a further puncture.

9   If the inner tube has been patched on a number of past occasions, or if there is a tear or large hole, it is preferable to discard it and fit a replacement. Sudden deflation may cause an accident, particularly if it occurs with the front wheel.

10   To replace the tyre, inflate the inner tube sufficiently for it to assume a circular shape but only just. Then push it into the tyre so that it is enclosed completely. Lay the tyre on the wheel at an angle and insert the valve through the rim tape and the hole in the wheel rim. Attach the locking cap on the first few threads, sufficient to hold the valve captive in its correct location.

11   Starting at the point furthest from the valve, push the tyre bead over the edge of the wheel rim until it is located in the central well. Continue to work around the tyre in this fashion until the whole of one side of the tyre is on the rim. It may be necessary to use a tyre lever during the final stages.

12   Make sure there is no pull on the tyre valve and again commencing with the area furthest from the valve, ease the other bead of the tyre over the edge of the rim. Finish with the area close to the valve, pushing the valve up into the tyre until the locking cap touches the rim. This will ensure the inner tube is not trapped when the last section of the bead is edged over the rim with a tyre lever.

13   Check that the inner tube is not trapped at any point. Re-inflate the inner tube, and check that the tyre is seating correctly around the wheel rim. There should be a thin rib moulded around the wall of the tyre on both sides, which should be equidistant from the wheel rim at all points. If the tyre is unevenly located on the rim, try bouncing the wheel when the tyre is at the recommended pressure. It is probable that one of the beads has not pulled clear of the centre well.

14   Always run the tyres at the recommended pressures and never under or over-inflate. The correct pressures for solo use are given in the Specifications Section of this Chapter. If a pillion passenger is carried, increase the rear tyre pressure only.

15   Tyre replacement is aided by dusting the side walls, particularly in the vicinity of the beads, with a liberal coating of french chalk. Washing-up liquid can also be used to good effect, but this has the disadvantage of causing the inner surfaces of the wheel rim to rust.

16   Never replace the inner tube and tyre without the rim tape in position. If this precaution is overlooked there is a good chance of the ends of the spoke nipples chafing the inner tube and causing a crop of punctures.

17   Never fit a tyre that has a damaged tread or side walls. Apart from the legal aspects, there is a very great risk of a blow-out, which can have serious consequences on any two-wheel vehicle.

18   Tyre valves rarely give trouble, but it is always advisable to check whether the valve itself is leaking before removing the tyre. Do not forget to fit the dust cap, which forms an effective second seal.

18 Fault diagnosis: wheels, brakes and tyres

Symptom	Cause	Remedy
Handlebars oscillate at low speeds	Buckle or flat in wheel rim, most probably front wheel	Check rim alignment by spinning wheel. Correct by retensioning spokes or by having wheel rebuilt on new rim.
	Tyre not straight on rim	Check tyre alignment.
Machine lacks power and accelerates poorly	Brakes binding	Warm brake drums provide best evidence. Re-adjust brakes.
Brakes grab when applied gently	Ends of brake shoes not chamfered	Chamfer with file.
	Elliptical brake drum	Lightly skim in lathe (specialist attention needed).
Brake pull-off sluggish	Brake cam binding in housing	Free and grease.
	Weak brake shoe springs	Renew if springs not displaced.
Harsh transmission	Worn or badly adjusted chain	Adjust or renew as necessary.
	Hooked or badly worn sprockets	Renew as a pair.
Tyres wear more rapidly in middle of tread	Over inflation	Check pressures and run at recommended settings.
Tyres wear rapidly at outer edges of tread	Under inflation	Ditto.

Chapter 7 Electrical system

Contents

General description ... 1	Neutral indicator bulb: replacement ... 11
Flywheel generators: checking the output ... 2	Speedometer bulb: replacement ... 12
Battery: inspection and maintenance ... 3	Parking light bulb: replacement - C70 model only ... 13
Battery: charging procedure ... 4	Resistance unit: general description - C70 model only ... 14
Selenium rectifier: general description ... 5	Courtesy light bulb: replacement - C70 model only ... 15
Fuse: location and replacement ... 6	Horn: location and adjustment ... 16
Headlamp: replacing bulb and adjusting beam height ... 7	Ignition switch: general description ... 17
Stop and tail lamp: replacing the bulb ... 8	Lighting switch and indicator switch ... 18
Flashing indicators: replacement of bulbs ... 9	Wiring: layout and inspection ... 19
Flasher unit: location and replacement ... 10	Fault diagnosis: electrical system ... 20

Specifications

Battery

	C50	C70	C90
Type	Lead acid, 6 volt		
Make	Yuasa Denki	Yuasa Denki	Yuasa Denki
Capacity	4 amp hours	4 amp hours	6 amp hours
Earth lead	Negative	Negative	Negative
Fuse value	15 amp	15 amp	15 amp

Lighting*

Headlamp bulb	25/25 watts
Tail/stop light bulb	5/21 watts
Flashing indicator	18 watts
Neutral indicator	1.5 watts
Speedometer bulb	1.5 watts
Parking light	5 watts (C70 only)
Courtesy light	3 watts (C70 only)

*All bulbs rated at 6 volts

1 General description

Two types of 6 volt electrical system are fitted to the Honda models, depending on the specification of the model.

The C90 model has a flywheel a.c. generator of the rotating magnet type which gives a high output. The C50 and C70 models have a flywheel magneto generator that supplies the ignition system without dependence on the battery. Both systems contain provision for charging the battery and supplying the lighting and other electrical loads after converting the output to d.c. by means of a rectifier.

2 Flywheel generators: checking the output

As explained in Chapter 4.2, the output from either type of generator can be checked only with specialised test equipment of the multi-meter type. If the generator is suspect, it should be checked by either a Honda agent or an auto-electrical expert.

3 Battery: inspection and maintenance

1 Two types of Yuasa battery having different amp hour capacities are fitted. The C50 and C70 models use a 4 amp hour battery, whilst the C90 uses a 6 amp hour battery.

2 The transparent case of the battery allows the upper and lower levels of the electrolyte to be observed without need to remove the battery. Batteries of the lead/acid type are employed.

3.2 The acid level can be seen through the case

Chapter 7: Electrical system

3.7 A single bolt retains the battery and carrier to the frame

5.2 The selenium rectifier is mounted by the battery

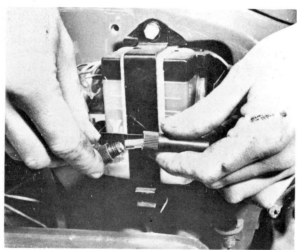
6.1 The fuse holder clips under the battery, and unscrews to release the fuse

Maintenance is normally limited to keeping the electrolyte level within the prescribed upper and lower limits and making sure that the vent tube is not blocked. The lead plates and their separators can also be seen through the transparent case, a further guide to the condition of the battery.

3 Unless acid is spilt, as may occur when the machine falls over, the electrolyte should always be topped up with distilled water, to restore the correct level. If acid is spilt on any part of the machine, it should be neutralised with an alkali such as washing soda and washed away with plenty of water, otherwise serious corrosion will occur. Top up with sulphuric acid of the correct specific gravity (1.260 - 1.280) only when spillage has occurred.

4 It is seldom practicable to repair a cracked case because the acid in the joint prevents the formation of an effective seal. It is always best to replace a cracked battery, especially in view of the corrosion that will be caused by the leakage of acid.

5 Never check the condition of a battery by shorting the terminals. The very heavy current flow resulting from this sudden discharge will cause the battery to overheat, with consequent damage to the plates and the compound they hold.

6 If the machine is laid up for any time, it is advisable to disconnect and remove the battery. It should not be allowed to discharge completely, otherwise sulphation is liable to occur, an irreversible change in the condition of the plates that will render the battery useless.

7 A single screw retains the battery carrier in position under the right-hand side cover.

4 Battery: charging procedure

1 Whilst the machine is running it is unlikely that the battery will require attention other than routine maintenance because the generator will keep it charged. However, if the machine is used for a succession of short journeys mainly during the hours of darkness when the lights are in full use, it is unlikely that the output from the generator will be able to keep pace with the heavy electrical demand. Under these circumstances it will be necessary to remove the battery from time to time, to have it recharged independently.

2 The normal charging rate for the two types of battery fitted to the Honda 50 cc and 90 cc models is 0.2 amps. A more rapid charge can be given in an emergency, in which case the charging rate can be raised to 0.6 - 1.0 amps. The higher charge rate should be avoided if possible because this will eventually shorten the working life of the battery.

3 When the battery has been removed from a machine that has been laid up, a 'refresher' charge should be given every six weeks if the battery is to be maintained in good condition.

5 Selenium rectifier: general description

1 The function of the selenium rectifier is to convert the a.c. produced by the generator to d.c. so that it can be used to charge the battery and operate the lighting circuit etc.
The C50 and C70 models have a half wave rectifier while the C90 model has a full wave rectifier.

2 The rectifier is located alongside the battery under the right-hand side cover. Apart from physical damage, the rectifier is unlikely to give trouble during normal service. It is not practicable to repair a damaged rectifier; replacement is the only satisfactory solution.

3 Damage to the rectifier is likely to occur, however, if the machine is run without the battery for any period of time. A high voltage will develop in the absence of any load on the coil, which will cause a reverse flow of current and consequent damage to the rectifier cells.

4 It is not possible to check whether the rectifier is functioning correctly without the appropriate test equipment. A Honda agent or an auto-electrical expert are best qualified to advise in such cases.

5 Do not loosen the rectifier locking nut (painted) or bend,

Chapter 7: Electrical system

cut, scratch or rotate the selenium wafers. Any such action will cause the electrode alloy coating to peel and destroy the working action.

6 Fuse: location and replacement

1 A fuse is incorporated in the electrical system to give protection from a sudden overload, such as may occur during a short circuit. The fuse is located within the fuse holder attached to the positive lead of the battery. All models have a 15 amp fuse.
2 If a fuse blows it should be replaced, after checking to ensure that no obvious short circuit has occurred. If the second fuse blows shortly afterwards, the electrical circuit should be checked in order to trace the fault.
3 When a fuse blows whilst running the machine and no spare is available, a 'get you home' dodge is to remove the blown fuse and wrap it in silver paper before replacing it in the fuse holder. The silver paper will restore the electrical continuity by bridging the broken fuse wire. This expedient should never be used if there is evidence of a short circuit, otherwise more serious damage will be caused. Replace the blown fuse at the earliest possible opportunity, to restore the full circuit protection.

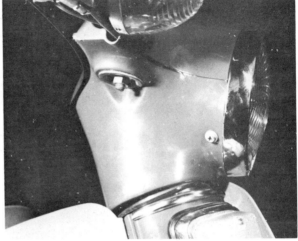

7.1a Remove the two headlamp retaining screws ...

7 Headlamp: replacing bulb and adjusting beam height

1 To remove the headlamp unit, detach the two bolts that hold the unit to the bottom half of the handlebar fairing and pull the unit clear. To remove the bulb, unhook the spring and pull the bulb holder free, then remove the bulb from the holder.
2 A double filament 25/25 watt bulb is fitted to take advantage of the high generator output.
3 It is not necessary to refocus the headlamp when a new bulb is fitted because the bulbs used are of the pre-focus type. To release the bulb holder from the reflector, pull from the locating flange.
4 Beam height is adjusted by means of the small screw on the lamp rim. Adjustments should always be made with the rider normally seated.

UK lighting regulations stipulate that the lighting system must be arranged so that the light will not dazzle a person standing in the same horizontal plane as the vehicle at a distance greater than 25 yards from the lamp, whose eye level is not less than 3 feet 6 inches above that plane. It is easy to approximate

7.1b ... and unclip the bulbholder to replace the bulb

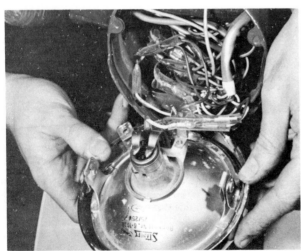

7.4 The reflector unit pivots for adjusting beam height

8.1 Slacken two screws to remove lens cover

Chapter 7: Electrical system

this setting by placing the machine 25 yards away from a wall, concentrating at the same height as the distance from the centre of the headlamp to the ground. The rider must be seated normally during this operation and also the pillion passenger, if one is carried regularly.

8 Stop and tail lamp: replacing bulb

1   The rear lamp has a twin filament bulb of the 5/21W type, to illuminate the rear of the machine and the rear number plate, and to give visual warning when the rear brake is applied. To gain access to the bulb, remove the two screws that retain the moulded plastic lens cover to the tail lamp assembly and remove the cover complete with sealing gasket.
2   If tail lamps keep blowing, suspect either vibration in the rear mudguard assembly, or a poor earth connection.
3   The stop lamp is operated by a stop lamp switch on the right-hand side of the machine, immediately above the rear brake pedal. It is connected to the pedal by a spring, which acts as the operating medium. The body of the stop lamp switch is threaded, so that a limited range of adjustment is provided to determine when the lamp will operate.

9 Flashing indicators: replacement of bulbs

1   The flashing indicators are located on the handlebar fairing. The rear-facing indicators are attached to the rear mudguard, below the seat or carrier.
2   In each case, access to the bulb is gained by removing the two screws and moulded plastic lens cover. Bulbs are rated at 18 watts.

10 Flasher unit: location and replacement

1   The flasher unit is located close to the battery, hanging vertically downwards from a single bolt fixing to the frame.
2   A series of audible clicks will be heard if the flasher unit is functioning correctly. If the unit malfunctions, the usual symptom is one initial flash before the unit goes dead. It will be necessary to replace the flasher unit complete if the fault cannot be attributed to either a burnt out indicator bulb or a blown fuse. Take care in handling the unit because it is easily damaged, if dropped.

11 Neutral indicator bulb: replacement

1   A neutral indicator light is incorporated in the speedometer, to show when the gear change lever is in neutral. A small contact in the gearbox selector drum provides the appropriate indication. Failure to indicate the selection of neutral can usually be attributed to a broken wire or a damaged contact.
2   The neutral indicator lamp is rated at 1.5W. It is a push fit into the rubber sleeve that holds it close to the green-coloured indicator glass.

12 Speedometer bulb: replacement

1   A 1.5W bulb is inserted from the bottom of the speedometer casing to illuminate the dial during the hours of darkness.

13 Parking light bulb: replacement - C70 model only

1   The parking light is mounted on the front cover, immediately below the headlight.
2   Access to the bulb is gained once the two screws and the moulded plastic cover have been removed. The bulb is rated at 5 watts.

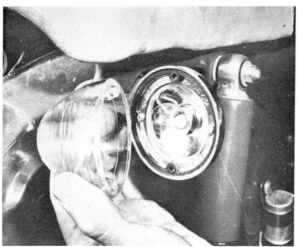
9.2 Remove two screws and plastic lens to replace bulb

10.1 The flasher unit is mounted by the battery

13.2 Remove two screws and the plastic lens to replace bulb

Chapter 7: Electrical system

14 Resistance unit: general description - C70 model only

1 A resistance unit is mounted alongside the air cleaner. Its function is to protect the parking light bulbs from a surge of power when the engine is started, with the parking lights on.
2 To remove the resistance unit, the legshield assembly must first be detached, the wires disconnected at the snap connectors and the central mounting nut and bolt removed.
3 A simple continuity test and a short circuit test are the only checks that can be applied, usually with a bulb and a battery.

15 Courtesy light bulb: replacement - C70 model only

1 The courtesy light is mounted alongside the ignition switch, protruding through the left-hand cover.
2 To replace the bulb, remove the cover, rotate the plastic button and pull it clear to expose the bulb. The bulb rating is 3 watts.

16 Horn: location and adjustment

1 The horn is suspended from the air cleaner hose, underneath the legshields.
2 There is means of adjusting the horn note at the rear of the horn body. If the horn note is weak, the adjusting screw should be turned anticlockwise to increase the volume.

17 Ignition switch: general description

1 The ignition switch is operated by a key, which cannot be removed when the ignition is switched on.
2 The number stamped on the key will match the number on the lock. This will aid obtaining a replacement key, if the original is lost.
3 The ignition key operates also the steering head lock.
4 It is not practicable to repair the switch if it malfunctions. It should be replaced with another lock and key to match.

18 Lighting switch and indicator switch

1 The switches are mounted on the handlebars and are not

14.1 The resistance unit is mounted behind the steering head

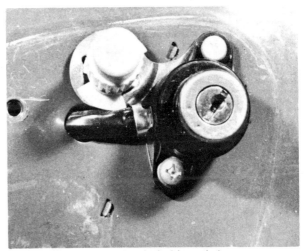
15.1 The courtesy light is by the ignition switch

15.2 The cover unscrews to allow access to the bulb

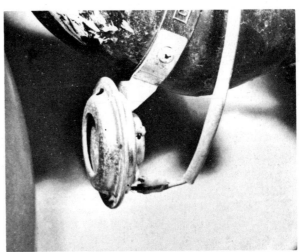
16.1 The horn is mounted on the inlet air hose

repairable. If any are faulty or damaged they must be renewed as a complete unit.
2 On no account oil the switches or the oil will spread across the internal contacts and form an effective insulator.

19 Wiring: layout and inspection

1 The wiring harness is colour-coded and will correspond with the accompanying wiring diagrams.
2 Visual inspection will show whether any breaks or frayed outer coverings are giving rise to short circuits. Another source of trouble may be the snap connectors, where the connector has not been pushed home fully in the outer housing.
3 Intermittent short circuits can often be traced to a chafed wire that passes through or close to a metal component, such as a frame member. Avoid tight bends in the wire or situations where the wire can become trapped between casings.

20 Fault diagnosis: electrical system

Symptom	Cause	Remedy
Complete electrical failure	Blown fuse	Locate fault and renew fuse.
	Broken wire from generator	Reconnect.
	Lighting switch faulty	Renew switch.
	Generator not charging	Check output.
Dim lights	Bad connections	Renovate, paying particular attention to earth connections.
Constantly 'blowing' bulbs	Vibration	Check bulb holders are secure.
	Poor earth connections	Renovate.
Indicators not working	Flat battery	Recharge battery and check generator output.
	Blown fuse	Check wiring before renewing 15 amp fuse.
	Isolated battery	Check battery connections. Clean any corrosion.
	Indicator switch faulty	Renew switch.

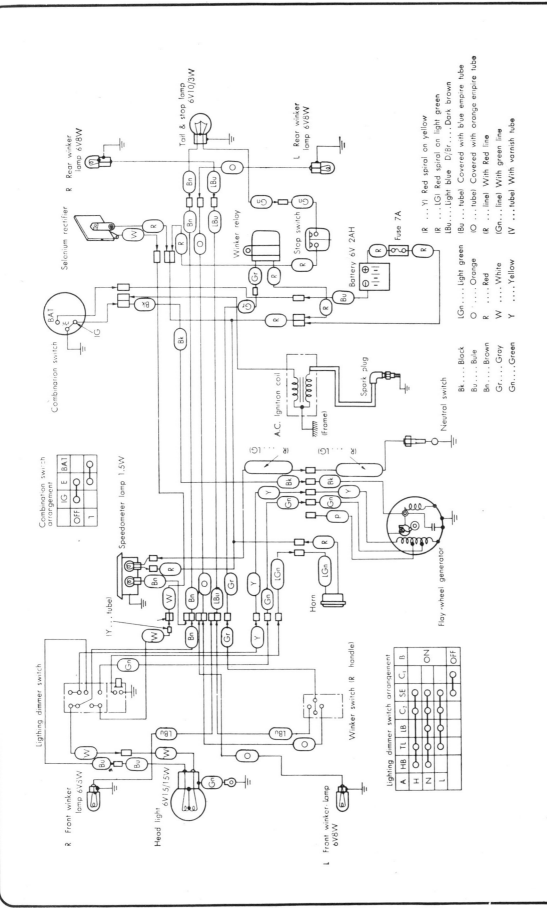

Fig. 7.1. Honda C50 Wiring Diagram

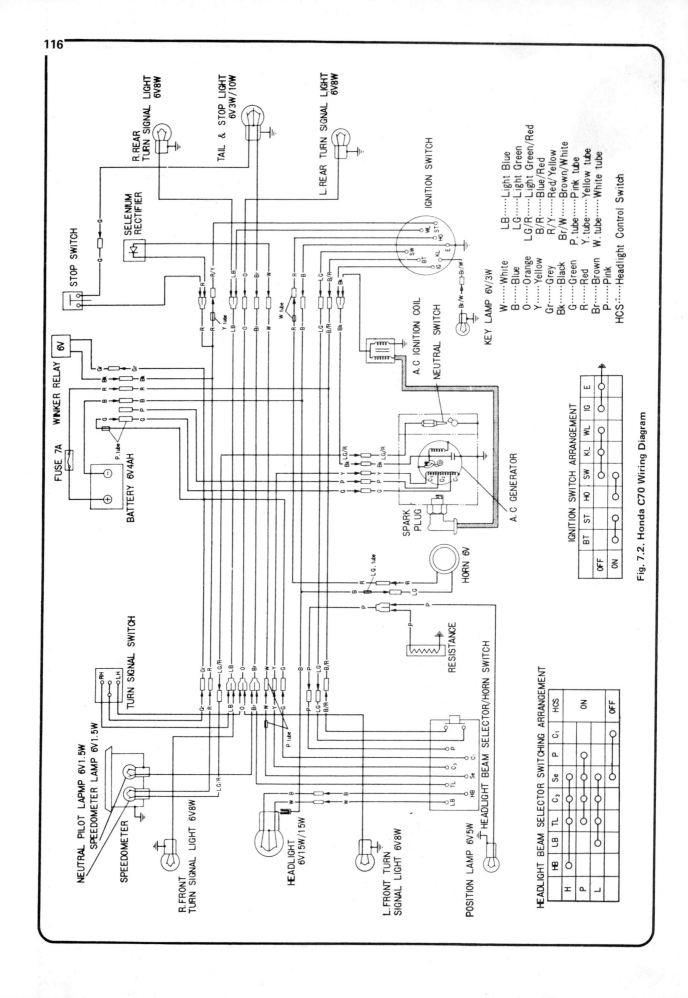

Fig. 7.2. Honda C70 Wiring Diagram

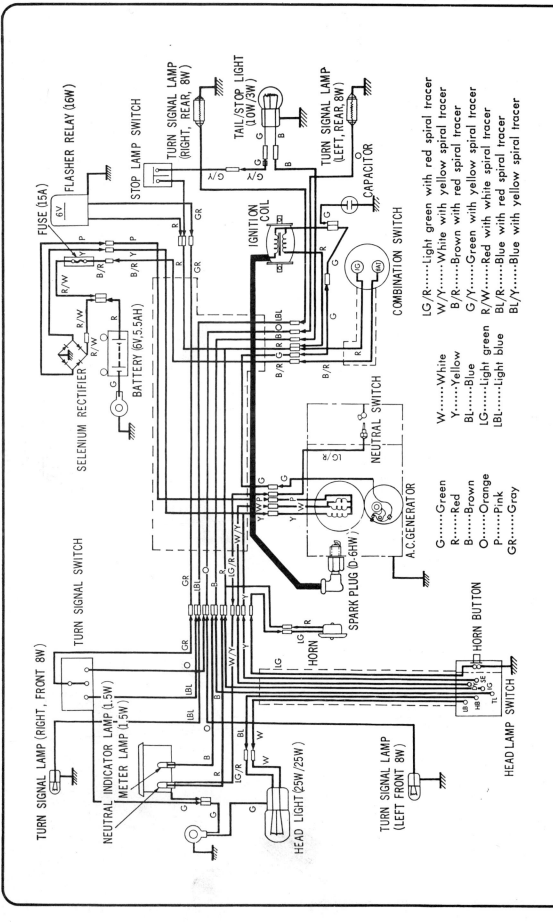

Fig. 7.3. Honda C90 Wiring Diagram

Metric conversion tables

Inches	Decimals	Millimetres	Millimetres to Inches		Inches to Millimetres	
			mm	Inches	Inches	mm
1/64	0.015625	0.3969	0.01	0.00039	0.001	0.0254
1/32	0.03125	0.7937	0.02	0.00079	0.002	0.0508
3/64	0.046875	1.1906	0.03	0.00118	0.003	0.0762
1/16	0.0625	1.5875	0.04	0.00157	0.004	0.1016
5/64	0.078125	1.9844	0.05	0.00197	0.005	0.1270
3/32	0.09375	2.3812	0.06	0.00236	0.006	0.1524
7/64	0.109375	2.7781	0.07	0.00276	0.007	0.1778
1/8	0.125	3.1750	0.08	0.00315	0.008	0.2032
9/64	0.140625	3.5719	0.09	0.00354	0.009	0.2286
5/32	0.15625	3.9687	0.1	0.00394	0.01	0.254
11/64	0.171875	4.3656	0.2	0.00787	0.02	0.508
3/16	0.1875	4.7625	0.3	0.1181	0.03	0.762
13/64	0.203125	5.1594	0.4	0.01575	0.04	1.016
7/32	0.21875	5.5562	0.5	0.01969	0.05	1.270
15/64	0.234275	5.9531	0.6	0.02362	0.06	1.524
1/4	0.25	6.3500	0.7	0.02756	0.07	1.778
17/64	0.265625	6.7469	0.8	0.3150	0.08	2.032
9/32	0.28125	7.1437	0.9	0.03543	0.09	2.286
19/64	0.296875	7.5406	1	0.03937	0.1	2.54
5/16	0.3125	7.9375	2	0.07874	0.2	5.08
21/64	0.328125	8.3344	3	0.11811	0.3	7.62
11/32	0.34375	8.7312	4	0.15748	0.4	10.16
23/64	0.359375	9.1281	5	0.19685	0.5	12.70
3/8	0.375	9.5250	6	0.23622	0.6	15.24
25/64	0.390625	9.9219	7	0.27559	0.7	17.78
13/32	0.40625	10.3187	8	0.31496	0.8	20.32
27/64	0.421875	10.7156	9	0.35433	0.9	22.86
7/16	0.4375	11.1125	10	0.39270	1	25.4
29/64	0.453125	11.5094	11	0.43307	2	50.8
15/32	0.46875	11.9062	12	0.47244	3	76.2
31/64	0.484375	12.3031	13	0.51181	4	101.6
1/2	0.5	12.7000	14	0.55118	5	127.0
33/64	0.515625	13.0969	15	0.59055	6	152.4
17/32	0.53125	13.4937	16	0.62992	7	177.8
35/64	0.546875	13.8906	17	0.66929	8	203.2
9/16	0.5625	14.2875	18	0.70866	9	228.6
37/64	0.578125	14.6844	19	0.74803	10	254.0
19/32	0.59375	15.0812	20	0.78740	11	279.4
39/64	0.609375	15.4781	21	0.82677	12	304.8
5/8	0.625	15.8750	22	0.86614	13	330.2
41/64	0.640625	16.2719	23	0.90551	14	355.6
21/32	0.65625	16.6687	24	0.94488	15	381.0
43/64	0.671875	17.0656	25	0.98425	16	406.4
11/16	0.6875	17.4625	26	1.02362	17	431.8
45/64	0.703125	17.8594	27	1.06299	18	457.2
23/32	0.71875	18.2562	28	1.10236	19	482.6
47/64	0.734375	18.6531	29	1.14173	20	508.0
3/4	0.75	19.0500	30	1.18110	21	533.4
49/64	0.765625	19.4469	31	1.22047	22	558.8
25/32	0.78125	19.8437	32	1.25984	23	584.2
51/64	0.796875	20.2406	33	1.29921	24	609.6
13/16	0.8125	20.6375	34	1.33858	25	635.0
53/64	0.828125	21.0344	35	1.37795	26	660.4
27/32	0.84375	21.4312	36	1.41732	27	685.8
55/64	0.859375	21.8281	37	1.4567	28	711.2
7/8	0.875	22.2250	38	1.4961	29	736.6
57/64	0.890625	22.6219	39	1.5354	30	762.0
29/32	0.90625	23.0187	40	1.5748	31	787.4
59/64	0.921875	23.4156	41	1.6142	32	812.8
15/16	0.9375	23.8125	42	1.6535	33	838.2
61/64	0.953125	24.2094	43	1.6929	34	863.6
31/32	0.96875	24.6062	44	1.7323	35	889.0
63/64	0.984375	25.0031	45	1.7717	46	914.4

Index

A

Adjustments:
  brake cables - 8
  carburettor - 9, 74
  clutch - 10, 65
  contact breaker - 82
  headlamp - 111
  ignition timing - 11
  tappets - 57
  throttle cable - 9
Air filter - 10, 75
Automatic advance unit - 84

B

Battery - 9, 109, 110
Bearings:
  front wheel - 13
  main - 42, 47
  rear wheel - 13
  steering head - 92
Brakes:
  adjustment - 105
  dismantling, examination and replacement - 99
  fault diagnosis - 108
  specifications - 98
Bulbs:
  courtesy - 113
  fault diagnosis - 114
  headlamp - 111
  neutral indicator - 112
  parking light - 112
  specifications - 109
  speedometer - 112
  stop and tail lamp - 112

C

Cables:
  lubrication - 8
  speedometer - 13, 96
  throttle - 13
Camchain - 38, 47
Camshaft - 26, 44, 45, 53
Capacities - 7
Carburettor:
  adjustment - 9, 74
  cleaning, examination and reassembly - 73
  dismantling - 71, 73
  fault diagnosis - 78
  removal - 70
  specifications - 67
Centre stand - 13, 93
Chain, final drive - 10
Cleaning - 10, 74, 75, 77, 96
Clutch:
  adjustment - 10, 65
  correct operation - 65
  dismantling - 61
  examination and renovation - 64, 65
  fault diagnosis - 66
  reassembly - 50, 65
  removal - 29
  specifications - 61
Coil - 82
  Condenser - 84
Contact breaker:
  adjustment - 82
  lubrication - 13
  removal, renovation and replacement - 82
Crankcase - 34, 49
Crankshaft - 34, 48
Cylinder barrel - 44, 54
Cylinder head - 20, 55

E

Electrical system:
  fault diagnosis - 114
  specifications - 109
Engine:
  dismantling - 19
  examination and renovation - 42
  fault diagnosis - 59, 60
  reassembly - 46, 47, 48
  removal - 17
  specifications - 15, 16
  starting and running - 59
Exhaust system - 75

F

Fault diagnosis:
  clutch - 66
  electrical system - 114
  engine - 59, 60
  frame and forks - 97
  fuel system - 78
  gearbox - 60
  ignition system - 86
  lighting system - 114
  lubrication system - 78
  wheels, brakes and tyres - 108
Flasher unit - 112
Flywheel generator - 82, 109
Final drive chain - 10, 106
Final drive sprocket - 13, 34, 49
Footrests - 96
Frame:
  fault diagnosis - 97
  inspection and renovation - 93
  specifications - 87
Front forks:
  dismantling - 89
  fault diagnosis - 97
  removal - 87
  specifications - 87
Fuel system:
  fault diagnosis - 78
  specifications - 67
Fuel tank - 68
Fuel tap - 68
Fuse - 111

G

Gearbox:
  dismantling - 19
  fault diagnosis - 60
  removal - 17
Gearchange mechanism - 30, 45, 49
Gear selector drum - 38, 45, 47
Generator - 19, 58, 82, 109

H

Headlamp - 111
Horn - 113

Index

I

Indicators, flashing - 38, 47, 112
Ignition:
    automatic advance - 84
    coil - 82
    condenser - 84
    contact breaker - 13, 82
    fault diagnosis - 86
    generator - 19, 58, 82
    specifications - 79
    switch - 113
    timing - 11, 84

K

Kickstart - 34, 38, 45, 48, 49

L

Link forks - 13
Lubricants, recommended - 13
Lubrication system - 67, 75, 78

M

Main bearings - 42, 47
Maintenance, routine - 7

O

Oil filter - 26, 50, 77
Oil pump - 38, 42, 47, 77, 78
Oil seals - 42, 47
Ordering spare parts - 6

P

Petrol pipe - 68
Petrol tank - 68
Petrol tap - 68
Piston - 25, 44, 54
Piston rings - 25, 44
Primary drive - 29, 45, 50

R

Rear suspension units - 93
Recommended lubricants - 13
Rectifier - 110
Resistance unit - 113
Routine maintenance - 7
Running in - 59

S

Seat - 96
Shock absorber - 105
Spare parts, ordering - 6
Spark plugs - 9, 85, 86
Specifications:
    brakes - 98
    clutch - 61
    electrical system - 109
    engine - 15, 16
    frame and forks - 87
    fuel system - 67
    general - 6
    ignition system - 79
    lubrication system - 67
    tyres - 98
    wheels - 98
Speedometer - 13, 96, 103
Sprockets, final drive - 13, 34, 49
Steering head bearings - 92
Steering head lock - 93
Suspension:
    fault diagnosis - 97
    specifications - 87
    units - 93
Swinging arm rear fork - 93
Switches - 38, 47, 113

T

Tail lamp - 112
Throttle cable - 9, 13
Timing:
    ignition - 11
    valves - 55
Tools - 7, 14
Tyres:
    condition - 10
    fault diagnosis - 108
    pressures - 7
    removal and replacement - 106
    specifications - 98

V

Valves - 26, 44, 53, 55

W

Wheel:
    bearings - 13, 101, 103
    examination and renovation - 98
    front - 98
    rear - 103
    reassembly - 105
    removal - 99, 103
    replacement - 103
    shock absorber - 105
    sprocket - 103
Wheels:
    fault diagnosis - 108
    specification - 98
Wiring:
    diagrams - 115, 116, 117
    layout and inspection - 114
Working conditions - 14

**Printed by
Haynes Publishing Group
Sparkford Yeovil Somerset
England**